# NORTH AMERICAN BIG-GAME ANIMALS

*text by*
## Byron W. Dalrymple

*photos by*
## Erwin Bauer

**Published by OUTDOOR LIFE BOOKS**

**Distributed to the trade by STACKPOLE BOOKS**

*Copyright © 1978 by Byron W. Dalrymple*
*Copyright © 1985 by Erwin A. Bauer*

*Published by*

> *Outdoor Life Books*
> *Times Mirror Magazines, Inc.*
> *380 Madison Avenue*
> *New York, NY 10017*

*Distributed to the trade by*

> *Stackpole Books*
> *Cameron & Kelker Sts.*
> *Harrisburg, PA 17105*

*ISBN: 0-943822-56-4*

**Library of Congress Cataloging in Publication Data**

*Dalrymple, Byron W., 1914–*
  *North American big-game animals.*

  *1. Big game animals—North America.   2. Mammals*
*—North America.   I. Title.*
*QL715.D33   1985        599.7'097        85-11489*
*ISBN 0-943822-56-4*

*Manufactured in the United States of America*

# CONTENTS

# INTRODUCTION

This book is about the lives of the mammals of North America that over the centuries have commonly been called "big-game" animals—about where and how they live, their daily and seasonal routines of feeding and movement, their sex lives, the birth and development of their young, how they use their keen senses, what signs they leave indicative of their presence, and about their unique relationships with man, past and present.

This group of animals, which includes all of the larger land-based mammals of the continent, was selected not because they have long been called *game,* but because they are in general the most interesting, most abundant, most visible, their lives the most dramatic, and closely entwined for centuries with man's. They are the animals about which wildlife enthusiasts of all inclinations invariably seek to become better informed. Although no creature is unimportant to the large scheme, and every life form is linked to every other, it is this group of animals that is envisioned by most people when the phrase "our wildlife heritage" is used.

These creatures are indeed the important nucleus of that heritage. The word "game" which links them together obviously derives from the fact that from earliest times these animals were taken by man for food and hides, and in modern times particularly by the sport hunter. That link is an important one in our age, what with the new spirit of concern for the environment and for wildlife.

No one can possibly know all there is to know about the animals even of one county of a single state, let alone all those of the entire continent. Even narrowing it down to the so-called big-game animals, the scope of endlessly accrued knowledge is vast and all but unmanageable—and still there are empty spots. For over half a century I have been an observer of the animals selected here, and for more than half that time a writer about them. Their lives, and our human relations with them, are endlessly intriguing studies. I have attempted to gather here knowledge gained from my own observation plus much more from research of the studies of others.

Today we are all slowly learning that the preservation of these animals is important not just to hunters, but to all of us. Their continued existence in healthy numbers illimitably benefits both practically and aesthetically the total environment, of which we too are simply a part, although an overwhelmingly disproportionate part so far as our influence upon the lives of other creatures is concerned.

There is no question that in the past, from the days of the white man's first settlement, man the hunter has committed many excesses. Before him, the Indian, invariably shown nowadays erroneously as a thoughtful conservationist, also in numerous instances pressured these animals to excess, by no means always taking just what were needed. However, Indian population was not large enough—as ours is—to create many problems. Hide hunters, meat hunters during days of early exploration and settlement did awesome damage. There was no regard for what we now know as "game management."

Fortunately, most of the excesses were recognized in time. It was these—the disregard for conservation—that finally focused attention on the need for restraint and order in our relationships with these animals. The hunter came to understand that rules had to be made, and obeyed. And from these first stirrings the intricate and fascinating science of the management of wild animals was born. Today this is often called "game" management, but the fact is that science is being applied steadily more and more to all animals.

The benefits of concentrated management are many, and important to all of us. Deer, down to a half million animals years ago, and entirely exterminated in state after state, now number in the many millions. Elk and antelope, brought to the brink of extinction by hide and meat hunting, have been astutely managed into stable abundance in every expanse of habitat suitable to each on the continent.

Very slowly we are beginning to realize—the hunter who coined the term "game" for these creatures, the non-hunter who enjoys observing and perhaps photographing them, and the anti-hunter—that all have a common goal: to keep these animals as abundant as available habitat will permit. Particularly concentrated attention is being given to those few of the larger predators which appear threatened or endangered.

We are learning, even though perhaps too slowly, that *living room* is the key to the abundance—the very existence—of these animals. Some, to be sure, like the whitetail deer, adapt rather readily to man's pressures and learn to live on the very fringes of dense human settlement. Others cannot make do without true wilderness. An end to the incessant destruction and degradation of habitats suitable for these animals must be the chief concern of all those, regardless of their views for or against hunting, who wish to keep these creatures among us in abundance. The crucial matter is preservation and improvement of remaining habitat. No longer is the so-called "balance of nature" possible, if indeed it ever was. Man long ago irrevocably upset the balance, but in his current attention to scientific management as a substitute, he has been and continues to be tremendously successful, and indeed can be even more so in the future.

Perhaps in the long view it will turn out to be true, curiously, that past excesses sparked our learning. A few excesses do indeed continue, but the effort among hunters themselves and law enforcement personnel to stop or control these is truly prodigious, and will, it is hoped, succeed. Perhaps in the future wildlife enthusiasts, regardless of diverse views, will join forces, realizing that in truth all have the same goals. If this book can lead readers to a better understanding of the lives of the so-called game animals, their ways, their needs, their present-day relationships with man and his activities, it may offer at least a small service in bringing together in concentrated common effort all those, of whatever bent, who wish them well.

*Byron W. Dalrymple*
Kerrville, Texas

# COMPARISON OF TRACKS

## Hoofed Animals

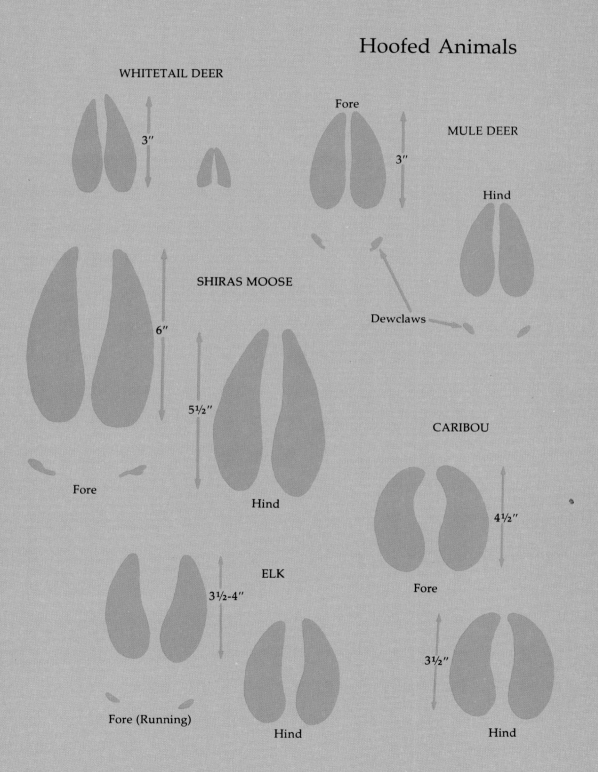

WHITETAIL DEER

3″

Fore

MULE DEER

3″

Hind

SHIRAS MOOSE

6″

5½″

Dewclaws

Fore

Hind

CARIBOU

4½″

Fore

ELK

3½-4″

Fore (Running)

Hind

3½″

Hind

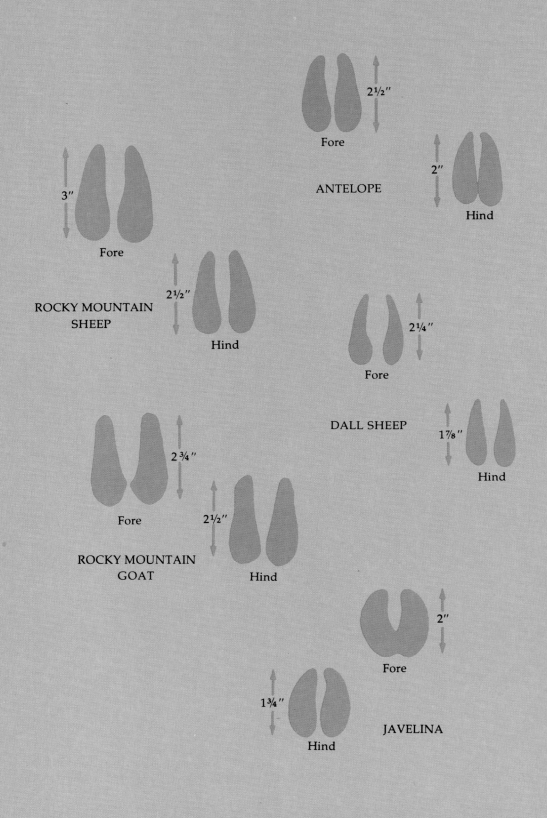

ANTELOPE

Fore    2½"

Hind    2"

ROCKY MOUNTAIN SHEEP

Fore    3"

Hind    2½"

DALL SHEEP

Fore    2¼"

Hind    1⅞"

ROCKY MOUNTAIN GOAT

Fore    2¾"

Hind    2½"

JAVELINA

Fore    2"

Hind    1¾"

# Bears

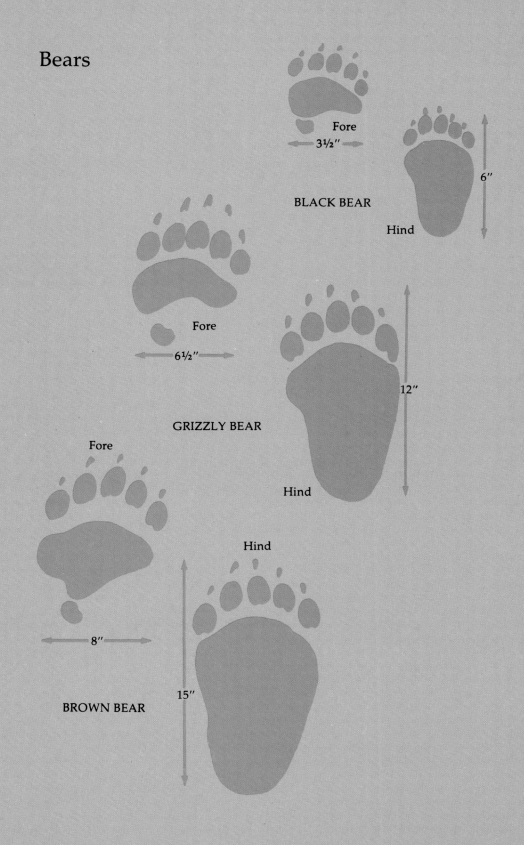

Fore
3½″
BLACK BEAR
6″
Hind

Fore
6½″
GRIZZLY BEAR
12″
Hind

Fore
Hind
8″
15″
BROWN BEAR

# NORTH AMERICAN BIG-GAME ANIMALS

# Whitetail Deer

*Odocoileus virginianus*

**T**he whitetail deer might well be designated the official antlered animal of North America. It is a true native; there are few theories that like some of our other animals it may have crossed a land bridge from Asia millions of years ago. It is of all large animals on this continent the most popular with both hunter and non-hunter. For all practical purposes, the word "deer" means whitetail deer to almost everyone. Paintings and drawings by tens of thousands of this trim creature, possibly the most handsome of all our larger mammals, have graced calendars, greeting cards, magazines, and books decade after decade. The whitetail is an inseparable part of the continent's culture and heritage.

Presently it is the most numerous of our larger animals. No one can be certain precisely how many whitetails there are, but estimates place their numbers at probably somewhere from 8 to 15 million. The range is enormous, covering much of southern Canada and most of the entire lower-48 states except two or three in the west, and reaching south throughout Central America.

Paradoxically, this shyest, most nervous, and wariest of all our deer is the one that was

*Although whitetail deer were abundant over three centuries ago when North America was first colonized, they were not as plentiful as today. The cutting of climax forests improved deer habitat immensely.*

able somehow to adapt to and cope with man's progress and to thrive with the thrust of civilization and progress all around it. In fact, the whitetail has proved to be one of the most successful animal colonizers of all time. It abounds in the big woods of northern Maine, and in the deep saw grass and hammock swamps of Florida. It thrives in the farmlands of southern Saskatchewan and grows so fat it can barely waddle in immense corn fields of the Dakotas. It is thoroughly at home in the cactus and thornbrush of southern Texas and on into Mexico, as well as in the forested areas of eastern Oregon and Washington.

Certainly the whitetail was abundant when the colonists first settled at Jamestown. But it may not have been as plentiful throughout the continent as it is today. Cutting of the climax forest and the forming of "edges" as agriculture and logging spread across the country enhanced whitetail habitat immensely. Yet this deer was to come close to extermination before it learned to live with its new neighbor, the white man, and finally to respond to his expert management. The story of the whitetail in this century is one of the great game-management stories of all time.

Early settlers utilized the whitetail for meat and hides, just as the Indians had but increasingly in far greater quantity. Need turned to greed when it was realized that both meat and hides were a valuable commodity. Hundreds of thousands of whitetail hides were shipped to England and Europe yearly, and matching numbers of barrels of venison went by ship or were utilized in budding American cities. Market hunting through the nineteenth century brought the whitetail population down to only a scant 250,000 to 500,000 animals. Commercial hunting stopped because there was no longer profit in it; deer were too scarce.

More important, however, the sport hunter now entered the scene. In some states—even Virginia, from which the whitetail had been named as the "type" species (the type from which all other whitetails would be judged in naming subspecies)—the whitetail was either totally or almost extinct. Setting of hunting regulations, and finally the science of deer management, which in the beginning employed trapping and transplanting of wild deer to ancestral ranges from which they had disappeared, has brought the whitetail to abundance or at least to stability on practically all suitable ranges on the continent. And in the meantime the animals themselves have become so adept at sharing man's habitat that in numerous instances deer live within city limits, although keeping their lives for the most part wholly private.

It is an odd commentary on whitetail abundance that Pennsylvania, one of the top whitetail states for numbers of deer, and for hunting, early in this century had almost none. Today Pennsylvania hunters annually harvest well over 100,000 on the average—and an astonishing and deplorable total of an additional 25,000 or more are killed annually on Pennsylvania highways by automobiles! In Texas, where the whitetail was considered on the way to extirpation at the turn of this century, it is possible to observe in the southcentral area as many as a hundred grazing during late afternoon on a small green-sprouting winter oat patch in early fall. The Texas herd, largest in the nation, is estimated at well over 3 million!

Generally speaking, whitetails of northern latitudes are larger, and darker, than their southern relatives. Maine checks in numerous 300-pounders (field-dressed) each season. Exceptional—possibly freakish—specimens have been taken much larger: a 440-pound buck in Iowa; two of 481 and 491 from Wisconsin; an astonishing 511-pound buck from Minnesota. Of the thirty subspecies recognized from the continent, seventeen of them are above the Mexican border. The majority of the subspecies especially within U.S. and Canadian borders, however, mean little today because most have been so diluted by transplants and interbreeding with others that few of the strains

*Showing alarm with his namesake, a whitetail buck displays the snowy underside of his tail. Whitetails also use their tails to signal "all clear."*

are pure. Further, a number have, or had, exceedingly limited ranges. Several, for example, are limited to individual islands off the Atlantic and Gulf coasts, and have developed specialized physical characteristics only because of their isolation.

The big northern woodland whitetail of New England and the Great Lakes region, *Odocoileus virginianus borealis,* is undoubtedly a pure race over much of its northernmost range. The same is true of the often larger Dakota whitetail, *O. v. dacotensis,* of Canada's Prairie Provinces, the Dakotas, eastern Montana, and Wyoming. It may surprise many

whitetail enthusiasts to know that in the official (Boone & Crockett Club) record book there are more whitetails from west of the Mississippi River by far than from east of it in what most hunters generally consider the prime whitetail country. Another surprise is that Saskatchewan has put far and away the most heads into the book.

Another large subspecies, the Kansas whitetail, *O. v. macrourus,* has been intergraded it is believed to extinction. However, the big northwest whitetail, *O. v. ochrourus,* of eastern Washington, Oregon, and into Idaho and bordering Rockies states, is probably pure in areas where it cannot intergrade. The handsome Columbian whitetail, *O. v. leucurus,* of small parts of coastal Washington and Oregon, is now a severely endangered subspecies, not hunted. Changes in land use have all but done it in. The tiny Florida Key whitetail, fully protected and with its own refuge, a deer which seldom weighs more than 75 pounds, is interesting because of its diminutive size and the fact that a special refuge saved it from extinction.

The one most important subspecies so far as hunters and most observers are concerned is the Coues or Arizona whitetail, *O. v. couesi.* This is a small, gray deer—mature bucks seldom weigh over 100 pounds—of the grass and oak and juniper of southeastern Arizona and southwestern New Mexico, most abundant at around 6000 feet altitude. Because of its isolation from others, this one is recognized separately in the record book, and is the only subspecies so recognized. A second subspecies of diminutive stature living high up in small, individual mountain ranges of the Big Bend Country of western Texas and on across into Mexico is the Carmen Mountains whitetail, *O. v. carminis.* It is an unusual and handsome trophy indeed, but intergrades at lower altitudes on its home range with the large Texas whitetail, *O. v. texanus,* which is also present in the general area.

Except in the case of a few subspecies such as the Columbian, the whitetail deer is presently in no danger whatever. Its natural enemies have been so severely controlled that predation is a negligible influence except in local noncontinuing instances. Although man—the hunter—may be classed as a predator, he is actually a key to proper management, not a danger as relates to hunting. The worst enemies of the whitetail deer are the expansion of industry and human population, which incessantly snip away at available habitat. Some estimates claim that an average of 3000 acres per day of possible whitetail living room

are usurped—by roads, mining, lumbering, oil exploration, urban sprawl, and changing agricultural land use. That's over a million acres a year!

Starvation and diseases related to malnutrition cause the greatest number of whitetail deaths. Occasionally these dieoffs are massive. And usually they are caused by overpopulation on certain ranges. The whitetail is a prolific creature, capable of doubling its numbers each year. Drought and severe winters when food supplies give out often cause these local debacles. The science of whitetail management, however, has been brought to a high art. Hunter quotas for antlerless deer are a tool used to keep herds tailored to their available food supply. Because of meticulous management, the handsome whitetail with its broad white flag waving as it bounds away will unquestionably be present in abundance over a vast area throughout the foreseeable future, for hunter and non-hunter alike to enjoy.

## Habitat

Trying to describe the habitat preferred by whitetail deer is like trying to be understood by several nationalities while speaking a single language. What an experienced New England deer hunter recognizes as perfect deer country is totally unrelated to what a hunter from near the Mexican border or from the land's-end swamps of southern Louisiana sees. As noted previously, the whitetail has been able to adapt to very different habitats, and to thrive in them. Yet the astute observer will soon perceive that the basics of habitat are easily related no matter how dissimilar the terrain may appear.

This deer is by personality an animal that must have numerous hiding places. Yet it never reaches high population levels in climax forest—that is, forests that are mature and thus have very little understory. It is commonly spoken of as an "edge" animal. Let's say there is an expanse of mixed forest, some conifers,

# THE WHITETAIL DEER

**Color:** Differs in shading locally, among subspecies, and seasonally; in general brown to gray-brown winter, reddish-brown summer; darker along back, shading lighter down sides to white beneath; nose black with white band behind at either side; white circles around eyes, white inside ears, over chin; white throat patch; upper inside of legs white, outside brown; tail brown above, some subspecies reddish, with black center swath, snowy-white long hair beneath and over rump area covered when tail is down.

**Measurements, mature bucks:** National average overall length 5 to 6 feet; height at withers, 3 to 3½ feet, variable among subspecies.

**Weight, mature bucks:** National average 125 to 160 pounds; highly variable locally and among subspecies, from 75 to 300-plus pounds.

**Antlers:** Main beams curving up, back, then well forward; unbranched points rise from main beams; a brow tine near the front base of each antler; total number of points differs (six, eight, ten, twelve) with age and vigor.

**Does:** Smaller by an average one-fourth; less blocky build.

**General attributes:** Graceful appearance; superb agility; nervous, extremely wary personality; unbranched antler tines except in nontypical or "freak" instances; white tail (beneath) raised upright and waving side to side when animal is startled and fleeing; animal actually smaller in height and length than most observers envision it, thus able to conceal itself in cover lower than might seem possible.

## *Range of the Whitetail Deer*

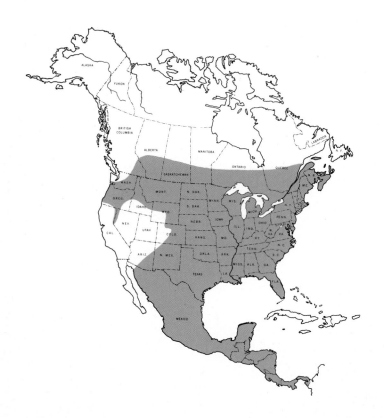

some hardwoods, cut by creeks, perhaps with a small lake or two thrown in, and here and there throughout a given expanse small to large openings. The openings may have scattered brush or shrubs, but they are ringed by denser mixed forest. The stream courses, the lake shores, the openings all encourage growth of forage, shrubs, and grasses. All these are edges of the forest, and it is these that are essential to the success of any whitetail deer herd.

In numerous locations where small farms checkerboard the countryside there are scores of privately owned woodlots, with perhaps larger tracts of state or national forest interspersed. Usually such part-farm-part-woods mixtures form optimum deer habitat. It is the brushy or wooded edges that furnish the most food for the deer; water is seldom a problem; and the thickets and denser forest serve as hiding and bedding places. A mule deer, an elk, or a caribou may bed down spang in the open where it can see far and catch a good breeze. Or a mule deer in particular may bed down high on a slope beside a single small bush or rock. The whitetail is totally different. It is secretive. When the terrain and growth allow it to both rest and feed without coming into the open, it will do so. Seldom will a mature whitetail cross an opening of any size. It will follow the edge, keeping to cover. In heavily settled states many a huge buck has lived undetected and unsuspected for some years in a near-town farm woodlot, feeding only at night.

In swamps of the south, deer find the all-

*The whitetail is the least inclined of native big game to travel far. But it is completely at home in its own environment. Standing still, this buck in a snowy northern woods is well camouflaged and difficult to spot, though hardly 100 feet away.*

important edges along the rim of the swamp itself, where large tree growth begins and where brushy thickets intergrade between forest and swamp. Lumbering operations along southern stream courses enhance whitetail opportunity. In the heavily agricultural areas of, for example, the Dakotas and Nebraska, there are willow-grown creek bottoms and gullies, plus windbreaks and even in season high stands of corn that serve as foraging and hiding places.

In the desert that the whitetail has been able to colonize, such as the so-called Brush Country of southern Texas, a rolling expanse of seemingly endless cactus and thornbrush, there are densely grown creek bottoms. These "creeks" deserve the name only during infrequent rains. But the water encourages thick vegetation. These are favorite hiding places for whitetails, which incidentally grow to trophy size in this rather unusual deer habitat. The grid of bulldozed ranch roads and seismograph trails from oil exploration throughout this large-ranch cattle country forms needed edges. So do openings where cattle ranchers clear the thornbrush so grass will grow.

Thus, you see, the habitat of the whitetail deer cannot be precisely detailed, as it can for moose, or elk, or pronghorn. It is highly and most interestingly diverse, in looks, altitude, latitude, vegetation growth, climate, and aridity or lack of it. Yet the fundamentals—dense thickets in which to hide and move about, edges to furnish much of the food—are in each case abundantly present.

## Feeding

Just as the nature of the habitat differs so broadly over the vast domain of the whitetail, so too must the varieties of plants, shrubs, and trees available in a given region. It is amusing, for instance, for a southwestern whitetail hunter to hear a New England deer hunter extol the value of an old apple orchard as a magnet for drawing deer, and telling how apple scent should be used by all deer hunters to overpower the human smell. To a whitetail on the Mexican border, which has never seen or tasted an apple, the smell might be frightening. There are in fact instances of men holding a deer-hunting lease in that country putting out bushels of corn in feeders to help keep the deer fat and in residence—only to discover that the animals wouldn't touch it. They had never tasted corn, didn't know it was fit to eat, yet they avidly gobbled up huajillo brush, a standard item of desert diet, which no doubt no Vermont whitetail would touch, given the opportunity.

Thus the hunter or wildlife enthusiast interested in whitetails should make a point of knowing which are favorite and abundantly available foods in his general region. These are only a few out of several hundred that deer are known to eat. Wherever oaks grow, for example, acorns are an extremely important item of diet. Buds and twigs of maple, sassafras, poplar, aspen, and birch are all staples. Sumac, beech leaves and mast, witch hazel, blueberry, blackberry, in the west mountain mahogany, varied ferns all over the continent, and wintergreen are all palatable to whitetails. Wherever they are found, wild grape vines, mulberry, basswood, serviceberry, wild rose, holly, and honeysuckle have a place in the diet. Wild clover, green grasses in spring, aquatic plants of lake and stream shores, and a broad variety of waste grain and vegetable crops as well as those upon which deer cause depredations are on the important list. So are chokecherry bushes, persimmon, greenbrier, and in specialized situations as in the southwest, yucca, huajillo, prickly pear cactus and its fruits, comal, ratama, and various tough shrubs often lumped under the name "buckbrush."

Wherever conifers grow, these are utilized, but chiefly in winter when other foods are scarce: jackpine in the Great Lakes region, juniper in the southwest, hemlock, white cedar,

*Whitetails are not frail, docile, Bambi-like creatures, as they are mistakenly depicted by some people. They are tough survivors. Does are quick to strike out at other does, especially if there is competition over space or food.*

fir, and others wherever they are available. These are not the most nutritious foods. Whitetails do not migrate to winter ranges, thus in snow country are inclined to "yard up" during severe winter in so-called cedar swamps and jackpine stands. Here they gain protection from weather, and feed on the browse. Many a northern "deer yard" has seen the browse line creep up and up until only the largest animals, rearing on hind legs, can reach any food at all. Then finally none is left and if deep snow continues, starvation wipes out scores of deer.

There have been many experiments in winter feeding. Hay has been used, but is not a proper diet. Occasionally volunteers working with game-department personnel go into a forest and cut poplar close to the yarded-up deer, so they can get to the twigs. But such measures are makeshift at best. Many whitetails starve or die of diseases caused by malnutrition during winter. This occurs in the south, too, when food runs low, and especially in snowless places like central Texas, where deer are extremely abundant and not all years offer substantial acorn crops to help pull the animals through to a newly green spring. Like all members of the family, whitetails have no upper front teeth. The size twigs they can handle are thus compared to their size, modest.

Deer do much feeding at night, especially

*Whitetails in winter often must rear on their hind legs to reach conifer browse.*

*Deer chew bushes almost to the snow line, until the branches are too thick to bite.*

on moonlit nights. Some observers believe whitetails prefer daytime feeding but are driven to night foraging when disturbed by the hunting season. This is a pat theory but probably only that. Conceivably under heavy pressure of the hunting season in a few areas deer may stay hidden all day. But the fact is, their eyes, which contain millions more light-gathering rod cells than the human eye, are tailored perfectly for night vision. Shine a light on a deer at night and its eyes glow; those of a human do not. It is a fact that regardless of hunting pressure deer move less as a rule in daytime during the bright moon phases than during the dark of the moon. Also, during hot

*Like most large mammals, whitetails are most visible when a blanket of snow covers the ground. If the snow becomes very deep and the food supply is depleted in late winter, survival becomes a desperate matter.*

weather the cool hours of dusk, night, and dawn are simply the most comfortable.

By and large, their chief feeding periods are from before dawn until several hours after, and again from late afternoon until dusk. The availability of food dictates how long it requires during each session to fill the paunch with 5 or 6 pounds of forage. In winter deer may be forced to nibble away much longer. They are also amazingly fussy and selective when enough food is present, passing up numerous edibles for those they most relish. Also, when there is a bonanza crop of an especially desirable food, whitetails, just like any other animal, including man, take the easy way,

moving as little as possible. During years when acorns literally by bushels are on the ground, whitetails may feed and bed down without ever showing themselves and without traveling more than a few yards a day. Life is then easy, and lazing away the day chewing cuds in safety and seclusion is far better than nervously searching the edges for scattered tidbits.

## Movements

An examination of the slender and almost dainty lower legs of the average whitetail deer would make it seem patently unsuited for swift travel through brush, over rocks and obstacle courses of downed timber in a forest. Yet this is the most agile of all our deer. It bounds at 30 miles per hour when really frightened, zig-zagging through the most tangled terrain, hurdling obstacles and even high bushes in graceful flying leaps. A whitetail can stand feeding beside an ordinary farm fence, raise its head to see something that startles it, and clear the fence from a standstill with an unbelievable flow of barely perceptible muscle motion. To successfully pen whitetails, an 8-foot fence is needed, and even then many go over the top.

Hunters have long said that a whitetail buck commonly runs away, flag flying, without even knowing for certain what startled it. And unlike its relative the mule deer, which may slow to a walk and start feeding again just over the ridge, or even pause on top to look back, the whitetail may clear several ridges before pausing. When running—bounding—this deer comes down first on its forefeet and the hind feet strike next, but past the fores. This typical running gait, plus the broad, raised, waving tail, plus the nonbranching points of the typical whitetail antlers are its chief easily recognized identification tags. The mule deer bounds so that the hind feet strike close to but behind the forefeet, its small-diameter stringlike tail (except in the Columbian blacktail) is seldom raised at all, and its antler points branch.

As an aside here it should be pointed out that both whitetails and mule deer bear nontypical or freakish antlers in many instances. The record book allows for this, having classes for both typical and nontypical. Certain areas of range invariably produce more, or less, of each class; mule deer are inclined to show nontypical antlers more often than whitetails. Genetic characteristics may be the reason for such antlers; mineral content of soil also may be a factor.

A whitetail deer walks most of the time when in motion, strolling slowly and with grace. It has a trotting gait also, raising a foot on opposite sides—perhaps left fore, right hind, then repeat oppositely—as it moves. The trot may be used when an animal is mildly concerned, or even when it is hurrying to join a companion, or to go to food or water. The bounding all-out gait is for escape, and it utilizes all the power of the animal, with the tremendous push from all fours that unwinds its bunched body into a sailing leap that may cover 15 to 25 feet. Whitetails are also adept swimmers, often entering large streams and drifting along as if for pure enjoyment. They also swim into lakes to escape predators such as feral dogs, or to visit islands. Does occasionally give birth to their fawns in such places, to add to seclusion and safety.

Whitetail deer utilize more "body language" movements than any other of the family. A feeding doe puts her head down, jerks it up five seconds later to look nervously around. She cocks her ears forward, staring at some imagined or real danger. She moves her head stiffly from side to side, the better to focus on the object or to make certain it isn't moving. She may stamp a front foot impatiently, possibly hoping to make the danger give itself away. This process, and the steady staring, may continue for a full minute or more. Then she gives a little switch of her tail. To a trained observer of whitetails, this is the "all clear." Immediately her attention will shift and her head lowers to feed again.

However, if the foot stamping continues she may begin to snort or "blow." This may be to prod the danger into making a show, or it may merely indicate growing nervousness. Now her tail begins to rise. At straight out—half mast—she may still eventually settle down. If it raises past that point and is cocked over to one side, she is about to run. Then it comes up straight, she whirls, snorting, and bounds away, flag moving from side to side.

If you are observing a feeding deer and it raises its head, cocks its ears, but not tensely, and is not looking at you, it may have heard or seen or sensed another deer. Many a deer gives away another's presence by this movement. Or during the rut a doe seems to glide out of a thicket into an opening. She pauses, looks back over her shoulder, ears moving. Then on she glides. Watch closely the thicket behind her! A buck will almost certainly soon appear. Such are examples of the body movements of the shy, wary whitetail that can impart to an observer the deer's intent or state of mind.

So far as travel as such is concerned, the whitetail is the least inclined. Of all our deer its home bailiwick is the smallest, and it will not leave the environs into which it was born, even to avoid starvation. On a range where food is genuinely abundant around the year, many whitetails do not utilize more than half a square mile of territory. On the average, a square mile contains nearly all, except that bucks in rut may break out of the home-territory barriers to trespass on the domains of neighboring bucks.

Some extremely interesting scientific studies have been done to monitor deer range. In one recent instance in Texas, a monitor station was set up and manned twenty-four hours a day at least two days each week. Deer of the vicinity were trapped, fitted with electronic collars, and released. Wherever they were, the monitoring station could pick up their individual signals at any time, know where each

*Whitetails are best seen early in the morning, anywhere in their vast range. Mostly they appear to feed around woodland openings, forest edges, or farm fields. In these areas they seem to be especially alert.*

deer was and which one it was. During the test none was ever outside a plotted square mile, and several never got more than a half-mile away.

Whitetails, living in such small home bailiwicks, come to know every stick, stone, tree, stump, and land contour intimately. That is why they are so uncannily able to escape pursuit, or to detect danger. This may be likened to a man driving a mile to work each day along the same route. Even the smallest new sign that goes up or the most minor chuckhole in the pavement is immediately noticed. In deer domain, anything that looks, sounds, or smells unfamiliar is suspect.

Biologists love to tell of the Michigan hunting experiment in the Upper Peninsula some years ago. A square mile of forest was fenced deerproof, and over a period of many weeks a tally on the deer population meticulously made, a task none too easy even for experts. Six top-notch deer hunters were then turned loose in the study plot. Nine bucks were present, plus thirty does. During four days of hunting this single square mile of what was home ground to the deer, no antlers were sighted. It took over fifty hunting hours before the first kill was made.

Whitetails make no migrations to winter range, as some mule deer do. As has been noted, they do yard up within their own territory during heavy snow. A mass migration to better feeding grounds might save them, but they seem unaware, and totally reluctant to leave home. Movements at any time to water or food or to bed down are all very brief. And routines are quite predictable. The resident deer follow certain trails, although these may shift seasonally as foraging or water supply dictates, or upon disturbance. A saddle in a ridge, a stream crossing, a path deer-made around the head of a steep canyon are all used day after day unless human intrusion forces the animals into new patterns.

These habits are advantageous to hunters,

of course, but not if the animals are pressured. Whitetails driven off a ridge where they are resting in cedars, let's say, may bound down and across an opening once or twice and thus offer a waiting hunter or photographer a shot. But after that they will quit the resting area to seek a new one. It should be stated that a bedding ground is never any permanent location. Whitetails do not return to use beds that served previously. They lie down wherever they happen to be and feel safe, although certain locations in the domain obviously serve best for the purpose and thus are used more than others.

All of the foregoing is by no means intended to indicate that the whitetail deer is a sedentary creature. Far from it. It is the most acutely tuned to its environment of all the deer. But it is also simply and strictly an incurable homebody.

## Breeding

The whitetail has scent glands between the two parts of the hoof on all four feet. It also has a musk gland called the metatarsal gland on the outside of each lower hind leg, and the larger tarsal gland on the inside of each hind leg at the hock. Scent from these glands is probably used by deer for keeping in touch with each other, and to leave a scent that may be followed by another deer. When fall and the rut arrive the tarsal gland becomes wet with its secretion and smells very strong.

By September the antlers of bucks have been rubbed clean and polished; they begin mock battles with small saplings and bushes. Whitetail bucks are not especially gregarious with each other at any season, as mule deer may be. But now as fall wears on they become more than ever loners. By mid-September in some latitudes and later on into November in others, the rut begins.

Each buck now stakes out his territory. He has made "rubs" on numerous saplings in his

*During the annual rut, or breeding season, this northern whitetail buck has detected the strong scent of a doe coming into estrus. Now it stands, lip curling, testing the wind for a better "fix" on the doe's location.*

bailiwick while polishing his antlers. Now at several locations he makes what hunters call a "scrape." He paws out a spot in soft earth, generally beneath an overhanging branch. He urinates in the scrape and reaches up to nuzzle, nibble, and touch his antlers against the overhanging branch. Just why he does this last is not well understood but it is commonly part of the breeding-season ritual.

These scrapes denote his territory. He also ranges outside his own area, and other bucks raid his. Thus battles ensue. Whitetails at times become vicious fighters. Injuries, and locked antlers, are fairly common. Whitetail bucks do not make any attempt to gather a harem. Does come in heat for a brief period, twenty-four

to thirty hours, each twenty-eight days. A doe in heat may find the scrape of a buck, urinate in it, and leave her musky trail as she moves away.

Does in heat are coquettish. When a buck pursues, the doe may run, usually circling. She pauses. The buck rushes up to her. She runs again. A buck whitetail during the peak of the rut is totally concentrated upon breeding. This is the only time of year when it loses all caution. Bucks often pass within a few feet of a hunter, mouth open, eyes glassy. They also become quite fearless and occasionally dangerous. When following a doe, the buck sometimes runs with its tail held rigidly straight out behind. This is a kind of body language which hunters or observers can use to advantage.

Whitetail deer are not vocal creatures. But now and then during the rut a buck may be heard to utter a low, rasping grunt, not loud but repeated several times, as it follows a doe. Often a second buck, eager for a doe, will trespass and try either to fight another for the doe with it, or else to run her off. Many years ago the art of rattling antlers for whitetails was conceived for use during the rut. It supposedly originated along the Mexican border of Texas and spread somewhat from there. But it may have been used by various Indian tribes elsewhere.

A hunter saws a pair of antlers from the skull, takes a stand, and clashes the antlers together. He rakes brush and gravel, all of it simulating the sound of two bucks fighting over a doe. The trick is effective, of course, only during breeding season. Some bucks rush in with eyes wild, ready to fight. Some sneak in as if intent on luring the presumed doe away while the simulated fight goes on. Young bucks are especially silly. One may run up to the antler-rattling hunter's hiding place, dart off, race back, always watching to make sure no mature buck is in sight. If one appears, it immediately retreats.

The neck of the whitetail buck, like that of most deer, is swollen during the rut. The rut period for each individual buck generally lasts about a month. Each services a doe during her brief period, and leaves her to find another. The tarsal musk and the urine leave readily followed trails. Occasionally two or more bucks may chase after a single doe. Then a fight is almost certain. Although the buck may service a number of does during the rut, the whitetail seldom becomes quite as bedraggled as bull elk do with their large harems.

If a doe fails to be bred during a first fall period, she will probably be bred the next month, or even the third. Thus some fawns appear later than others, and a few are seen each year with their spots still evident well along in fall. The overall period of the rut on the average is from October through December. It differs from place to place and especially season to season. Some breeding action may be evident some years as early as September and as late as late January.

During this period when the heavily haired tarsal glands are wet with musk, hunters sometimes cut them off the legs of a freshly killed animal and use them on a stand as an attractant, or to mask human scent. Just how effective this may be is questionable. However, many hunters slick away these skin glands

*During the rut, whitetail bucks are likely to engage in head-to-head combat to determine rank and to claim a doe. This sometimes results in locked antlers. Unless found and separated by man, one or both bucks may die.*

from the inside of the hind leg before dressing out the animals, to avoid getting any musk on the meat.

Once the rut is over the personality of the buck reverts to normal once again. The wide ranging that has occurred during this period ceases. Each now stays within its own range and is shy and secretive once more. Through December and January as breeding ceases for each individual buck, the antlers begin to loosen and are shed. From that time on through the winter the bucks usually consort with other deer of either sex. Not until the new antlers form do they begin to get the urge to stay apart.

For many years it was thought that only fully adult whitetails were capable of breeding. It is now known, however, that on optimum range a substantial number of doe fawns are

*A nursing fawn does not follow its mother when she browses. Instead the fawn stays behind and, if danger threatens, assumes this motionless posture.*

bred during their first fall, and have their first fawns when they are only a year old. Some bucks breed when they are long yearlings—into their second fall; the vigorous activity, however, begins for most in the third fall.

## Birth and Development

When spring is well along in the woodlands and ground cover green and high, the reddish-colored fawns, dappled with white spots, begin to appear. Each doe gives birth a bit over 6½ months from the time she was bred. Twin fawns are common among adult does, and triplets not rare. Yearling does seldom drop more than a single fawn.

There is no group fawning area. Each doe finds her own haven. Whitetail fawns are unable to stand when first born, and are very wobbly on their legs for a few days. At birth they weigh only 4 to 5 pounds. For at least the first couple of weeks of their lives, and sometimes longer, the fawns stay at or within a few steps of the spot where the mother has led them as soon as they can walk. This hiding place or nursery is shady, possibly in deep grass beneath bushes, where the youngsters will be comfortable and safe. They do not follow their mother around.

This often leads to a "lost" fawn being caught and taken home by some well-meaning tourist. The fawn is not lost. It stays put while the mother forages. She does not go far, and she comes back to the hideaway several times daily to allow the fawn to nurse. When a fawn is first able to move about over a small area around the hiding place, it will drop to the ground with head outstretched at any indication of danger. An immobile fawn in dappled light and shadow of cover is difficult to spot, and apparently has little odor. When it is around a month old it begins to follow its mother.

Now and then it mimics her, nibbling at a plant here and there. Presently it is taking for-

*Toward the end of spring, when ground cover is green and high, reddish-colored fawns, dappled with white, begin to appear. Each doe gives birth to one fawn or twins, rarely triplets, within six-and-a-half months after breeding in the late fall.*

age, and progressively becomes less dependent upon milk. By fall the spots and fawn color are replaced by the first grayish coat of winter. Fawns of either sex may stay with the mother on through the first winter, or they may not. Buck fawns are more likely to wander off on their own.

Throughout their lives, as mentioned earlier, whitetails are not especially vocal. Very young fawns bleat on occasion. Grown fawns in their first fall may utter this sound if they have strayed from their mother. An injured adult deer is capable of uttering a startlingly loud "blatt" or bawl. This is not a common sound. Rarely an injured mature buck utters a low, harsh, rasping cry. The whistle or snort of a disturbed whitetail is the sound most often heard.

While the fawns are weaning and losing their spots, the bucks are off by themselves, their new antlers bulging in velvet. There are many misconceptions about whitetail antlers. Many wildlife observers, especially hunters, still believe the deer adds a point on either side annually, and thus can be aged by counting points. This is untrue. Antler growth depends upon the quality of the basic habitat—that is, the soil. If it contains proper minerals, these are passed along in food and water, and antler growth will far exceed that on poorer soils.

Further, the type of season is all important. If winter forage is abundant, weather is not too severe, and there is a "good" spring, the bucks will be in excellent physical condition as the antlers are forming and this will be evident in heavy, dark, well-formed antlers in fall. In poor seasons antlers are invariably slender, smaller, pale, and sometimes ill-formed.

Ordinarily a spring fawn will be in fall what hunters term a "button buck." The antlers are mere nubs that barely break through the skin, if at all. As a long yearling the buck will probably be a spike—each antler a single spike from 3 to 5 inches long. However, under the best conditions it may be a forkhorn, or even more. On the average over most whitetail range, the males progress from button bucks to spikes to forkhorns—each antler simply forked and with no brow tine. The following fall it may have six points or eight points. But there is no rigid rule. An "eight-pointer" is a buck with brow tines, two points rising from the main beam of each antler, plus the point formed by each main beam.

Some whitetails at maturity never show more than eight points. Others, vigorous, well-fed animals, are ten-pointers, or more. By and large, typical antlers of ten points total are the standard for mature bucks in their fourth or fifth year and onward. A few develop twelve points or more. Most antlers with numerous points fall because of their unsymmetrical conformation into the nontypical class.

The soil quality of the range, which is responsible for abundant or poor forage, is partly responsible for antler development. Recent experiments show that a high-protein diet assists antler growth. Deer fed all they will eat of such a diet often skip the spike stage and become six-pointers or better their first year. However, parallel studies at the Kerr Management Area in Texas, continued over a period of six years, indicate that genetics is an even more powerful influence. Massive-antlered bucks beget the same. When high-protein diet and plenty of it, plus genetically superior deer are combined, a kind of "superdeer" is produced. Somewhat similar studies in Tennessee with hybridizing of whitetails and blacktails show that there are sex links in the genetic carry-over: a blacktail buck bred to a whitetail doe gets progeny with bifurcate antlers like the male parent, and vice versa.

Bucks in their fourth and fifth years usually are at peak, but if one remains exceptionally vigorous over the next couple of years, it is likely to have extremely heavy antlers. Old bucks past their prime may not even have antlers, or may grow gnarled, ill-formed spikes.

Although not common, the incidence of color variations in the whitetail is far from rare. True albinos seldom occur, but at least one instance is recorded of a doe in New York State that gave birth to several albino fawns, along with fawns of normal color. The curious aspect of these albinos was that the skin was pink, and one, a buck, grew antlers that in the velvet were pure white.

"Paint" whitetails—spotted with blotches of brown and white—show up annually in hunter harvests. Usually a certain area produces these with fair consistency over the years, indicating a genetic aberration in a race of that location. Such individuals are not albinos. On the 7500-acre Seneca Army Depot in New York State a small herd of pure-white mutations has evolved from a single buck seen there in 1957. These also are not albinos. During a recent count the strain had proved so dominant that

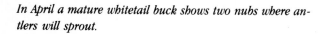

*In April a mature whitetail buck shows two nubs where antlers will sprout.*

*By late June the antlers have reached partial growth and are covered in a velvet-like substance that supplies nourishment.*

*In September the bucks rub the velvet from their full-grown antlers in readiness for the rut. This is an eight-pointer.*

*The whitetail's antlers are typified by a single main beam on each side with a number of tines, or points, growing off each beam.*

*Bucks are usually in their prime during the fourth or fifth years when antlers may reach their greatest size. This is a truly magnificent ten-point buck with almost perfectly symmetrical antlers. A hunter spending a lifetime in deer country would not be likely to spot a buck with antlers that could match these.*

*Bucks with slightly asymmetrical antlers are considered nontypical for trophy purposes. In many cases, this characteristic is hereditary. But the asymmetry and points can be far more freakish, often the result of damage to a growing antler, or other physical injury, or to a glandular problem.*

about a fifth of the deer on the depot were white. These deer are carefully protected. A few token hunts, meticulously supervised, have been allowed to keep the animals tailored to their range.

Melanism—a black phase, the opposite of albinism—apparently is much more uncommon among whitetails. No wholly melanistic specimens are known, but occasionally partial melanism appears. It has been recorded occasionally in the Adirondacks, and a portion of one county in the Texas Hill Country has turned up a number of very dark-coated bluish-black animals over recent years.

## Senses

The keenness of the sense of smell in the whitetail is legendary. The slightest whimsical air movement will waft scent of danger that is picked up at long distances even in the most meager amounts. Hunters report taking a stand in a tree and having whitetails walk right under them without recognizing their presence, yet deer 300 yards away snorted and fled as they came across the scent-drifting flow of a breeze so gentle the hunter was unaware of it.

Most animals that live in heavy cover are well equipped with highly developed scenting ability. They need it more here than, for example, does the antelope on wide-open plains. On days of high wind that gusts from varying directions, whitetails are especially nervous and easily spooked. They cannot keep a steady "scenting beam," and in addition their hearing is impaired by the noise around them.

By and large the whitetail depends first upon its nose, and then begins to focus the other senses to the degree the situation allows. Anything a whitetail hears or sees it immediately tries to smell. Scent is the real clincher of danger. A deer coming in to rattled antlers, for example, will circle if there is the slightest breeze. It may not be the least suspicious, but instinctively it tries to sniff out what's going on in order to authenticate what it has heard. Anyone who wishes to observe whitetails unaware must remember to keep even the slightest air movement in his favor.

Whitetail hearing is also acute. But it is secondary to scenting ability, because often the deer hears something and does not properly identify it without help of the nose. This is one reason why a whitetail will often race away at some slight sound and keep right on running over the next two ridges. It has heard something it could not check out with its nose. Conversely, experimenters have reported sitting well camouflaged in a blind above the ground and mumbling in low voice at deer down below. The deer might become nervous, but would soon also become accustomed.

One interesting aspect of whitetail hearing is that they seem to recognize the sounds made by another deer walking along, for example over rocks that clink. They've been observed merely cocking ears toward this sound, unconcerned. Yet a man making the same sound is instantly recognized as a disturbing element.

Whitetail eyesight is sharp for any movement. But it is not acute unless there is movement. A deer will recognize something amiss in its domain—a hunter hunched in camouflage, for example—but simply looking at it may be no more than mildly disturbed. The slightest movement, however, will put the animal to flight. Whitetails are colorblind, or nearly so, and thus colored clothing does not disturb them, unless in high contrast to the surroundings in the shade of gray they see. The eye placement and physical build of deer make it difficult for them to see above, and also at acute angles out to the side. The eyes are set chiefly for looking ahead and at right angles outward to the sides.

## Sign

Earlier the scrapes made by bucks during the rut were described. Rubs were also mentioned. The rub of a whitetail buck is usually made on a rough shrub, such as a small balsam in the north, a sapling pine or cedar in the south. It is simply a place where antlers have scraped bark from the sapling, and broken branches. Seldom are large trees used. Those an inch or less in diameter are most common. Where cattle and horses are pastured these domestic animals commonly rub their necks on larger trees, quite low to the ground. Occasionally a hunter mistakes these for deer sign. The size of the tree utilized is the clue.

The rub of a deer also is at head height of the animal or slightly lower. This means lower than a man's belt line, as a rule. In any given area bucks seem to select certain sapling or shrub varieties as favorites for rubbing. If one learns these, it helps locate animals. As an example, the green-barked, thorny retama of the southwest seems to attract whitetails in that region where small conifers are not present.

Over almost all of their range, deer tracks cannot possibly be confused with other tracks. They are totally unlike tracks of any domestic animals. Over the eastern half of the continent within whitetail range there are no other hoofed big-game animals. In some places of the west, whitetail and mule deer tracks may be confused. That makes little difference. Further, the terrain types favored by each help keep them fairly well separated. Galloping whitetails leave tracks with hind feet ahead of the forefeet, as noted earlier. Mule deer bound with hind-feet prints behind those of the forefeet. It is not really possible to distinguish between whitetail and mule deer tracks except when the animals are running.

Fawn tracks and javelina tracks might cause

*As a result of polishing their antlers and actually engaging in mock battles with shrubs and trees, bucks leave behind the scarred sign shown here. Bucks also rub their antlers on branches just above the ground scrapes they make with their hooves when staking out territory.*

confusion in the javelina's range. Fawn tracks, however, have sharply pointed toe imprints. Those of the javelina are blunt and rounded.

There has long been argument among hunters about how to tell buck tracks from doe tracks. The fact is, there is no sure way. A buck may or may not sink in soft earth more deeply. It depends upon his age and size. A buck, it is sometimes said, drags his feet more, showing drag marks in snow. All deer show drags in snow an inch or more in depth. Drags in a light skiff of snow may well indicate the track of a buck—or of a big old doe!

Deer droppings are another sign. Depending on the size of the deer, and the type of forage, these may be anywhere from a half-inch to over an inch in length. When the deer are eating soft summer food the droppings do not separate into individual pellets. Except in mule deer range droppings of whitetails cannot be readily confused with others. Occasionally rabbit droppings may cause a puzzle, but they should not because those of rabbits are round, not elongated.

Deer beds give some indication of the presence of animals on the range, but they are not purposely reused, so they are of interest only in passing. The only close-focus information the flattened grass or leaves or melted snow of deer beds will offer is a guess at what sexes made them. A group of three is almost certain to be left by a doe and two fawns. A single, large impression may be a big buck—or a big doe. Although whitetail bucks very occasionally wallow in mud of a scrape in which they have urinated, deer wallows are rare and this sign likewise.

## Hunting

Hunting whitetails is a tricky endeavor indeed because of their wariness. However, from 10 to 20 percent of hunters collect deer on the average range, and 50 to 75 percent are successful on ranges where the deer are ex-

*The art of rattling antlers was probably developed along the Texas-Mexico border for use during the rut. The ruse is so effective in bringing whitetail bucks closer to hunters that it is becoming popular elsewhere. With antlers sawed from a skull, a hunter clashes them together and intermittently rakes brush and gravel with them to simulate the sound of fighting bucks.*

tremely plentiful, which proves that the hunter can still outwit his quarry, if he uses craft.

Many volumes have been written on the how-to of whitetail hunting. It is an involved subject, with many facets. Basically, however, there are four main methods in use today: stand hunting, still hunting, driving, and calling. One has already been partially explained—the calling technique which involves rattling antlers during the rut. An adjunct to that kind of calling is the use of deer calls now marketed and quite popular. Full instructions

come with these. Don't expect miracles, however. Deer calls are interesting to experiment with, but hardly solve all hunting problems. Many hunters also nowadays add scents. Some supposedly attract deer, others serve to mask human scent. Scents are used by still hunters and stand hunters as well as when calling.

Stand hunting is undoubtedly the most popular over all whitetail range, possibly because it entails the least exertion for the hunter and avoids disturbing the deer. It is most successful early and late in the day, when deer move most. The hunter selects a stand, preferably one which places him above where deer may move, as on a ridgeside or a knoll. Because whitetails utilize cover and skirt the edges of openings, clinging to cover as much as possible, small openings scattered over the viewing area are preferred to large ones. Stands taken near a scrape during the rut or positioned to allow watching a well-used trail, a saddle between knolls, or a stream crossing all are possibilities. Any area where deer have been steadily utilizing any landscape feature will bear watching. In farm country deer habitually jump a particular low spot on a fence to and from woodlot and field. This makes a perfect spot to watch.

The stand is selected so that any breeze is either toward or angling from the front across. The hunter conceals himself, either sitting with back against a tree, or in a thicket from which he can see out plainly. Never should he be skylighted. Camouflage clothing and even camo grease paint for the face, or a headnet, help the stand hunter "disappear." He should remain immobile and silent. Binoculars are all but mandatory. In fact, for any kind of deer hunting they are invaluable. Where it is legal nowadays many hunters use a tree stand. In some areas they also use metal seats high on a tall tripod of steel poles. These are currently marketed. Such placements allow the hunter to see into and over dense cover, and place him where deer are not likely to see him.

The still hunter slowly prowls the terrain, watching for deer. Again, the breeze must be kept in one's favor. Camouflage clothing is an assist. But most of all, slow movement counts. This means much slower than most hunters move. The hunter who spends an hour moving 300 yards sees more, makes less noise, and is far more likely to be successful than one walking along at normal speed. The still hunter utilizes every bit of cover, and is ever alert and with rifle ready. Many a bedded buck has been collected by the expert prowler. And many a

deer feeding along and unaware has been taken by the still hunter who keeps a sharp lookout ahead, then makes a perfect stalk.

In both stand and still hunting, an intimate knowledge of the terrain and the habits of the local deer plus utilization of this knowledge to one's best advantage are what bring success. The same is true, of course, when making a drive. The whitetail drive is more popular in the east than elsewhere. It can be dangerous unless hunters are well disciplined. And it is seldom successful for the tyro because he does not know enough about the habits of the deer.

A drive is made by placing several hunters at strategic locations near which deer pushed out of a resting place are most likely to move. A wooded ridge, let's say, with under cover suitable for a bedding spot, may overlook a small field at its end, with cover around the field rim. Drivers moving through the cover along the ridge push the deer off it before them. The deer may race across the small opening, or run along its edges, offering opportunity to hidden hunters.

There are many variations, but in all the terrain is used to force the deer to go where the shooters wait. Whitetails, however, are masters at doubling back. Expert drivers always try to check out where the deer have been moving naturally. They have normal travel routes. A drive that gently pushes them without undue disturbance along a normal travel route has the best chance of success. In a few states and provinces driving deer with dogs is still legal. This rather specialized technique, used chiefly in the south over many years, is presently declining.

Many different rifle calibers are adequate for whitetails. One should certainly not go undergunned, but the heavy magnums are hardly necessary. Shotguns using rifled slugs are mandatory in numerous heavily populated areas, and of course the bow and arrow is nowadays the favorite of many deer hunters.

Even with the rather astonishing annual

harvest of whitetail deer, this much revered and stately creature has a bright future, given proper continuing management, for many years ahead. As stated earlier, hunting is in fact the prime tool of management to keep this animal, which has so brilliantly adapted to utilizing the fringes of civilization as well as wilder places, matched in abundance to available range. The whitetail has few serious enemies in the wild, none of any real consequence among wild predators. Some are killed by coyotes, and a few by bears, bobcats, and the occasional lion. The worst predator over much of the range is the domestic dog. And the worst killer of all in every well-settled area is the automobile.

Starvation and disease do take a heavy toll at times. But the whitetail is capable of an astonishing replacement whenever its numbers are depleted. The continuing and ever increasing loss of habitat because of man's so-called progress is the worst danger the whitetail faces. Yet even that, for the foreseeable future at least, seems incapable of diminishing whitetail numbers to a danger point.

# Mule Deer

*Odocoileus hemionus*

The mule deer is a true westerner with a personality radically different from its more numerous and wider-ranging relative, the whitetail. On occasion it has been called low in intelligence, but that is a gross misinterpretation. It may be more naive and less wary than the whitetail simply because it is a true creature of the wilderness. Although mule deer are forced in some instances to live on the fringes of civilization, they seldom do well under such circumstances, and seem unable to adjust as well as the whitetail to man's progress and presence.

Throughout the west the mule deer is as much revered by sportsmen and wildlife observers as the whitetail is eastward. It gives the appearance of a big, rough, burly character, and indeed it is. But it is also a placid personality, not given to jittering and racing off without knowing what frightened it. Its curiosity about strange sights in its domain often gets it into trouble as it stands to stare, bounces stifflegged with a pogo-stick gait a few steps, stares again, finally runs, then perhaps stops atop the first ridge to look back. Nonetheless, mature bucks know every inch of their mountainous bailiwicks and can be canny indeed,

*Mule deer are widely distributed in the West. Although in some instances they are forced to live on the fringes of civilization, they are seldom able to adjust to human presence. This buck has an especially high rack of antlers.*

perhaps not wary in the super-alert manner of the whitetail but simple, direct, and crafty at staying out of sight, slipping away unnoticed, or simply moving away from disturbance to more remote regions.

It has been estimated that there are possibly half as many mule deer as whitetails, perhaps more. Figures are set variably with a top guess of around 7,500,000. Although the range of the mule deer is not as great as that of the whitetail, it nonetheless stretches over a vast area, from southeastern Alaska where the Sitka blacktail subspecies lives to far down into Mexico. It is especially interesting to note that the type species, the Rocky Mountain mule deer, has the broadest distribution of any antlered or horned game animal on the continent, from slightly above the 60-degree parallel in the far north to about the 35th in the middle of Arizona and New Mexico.

Although the mule deer has been a remarkable colonizer over an immense north-south expanse, curiously it has never been able to extend its range eastward in the same proportion as the westward thrust of the whitetail. It has been found sparsely in western Minnesota, abundantly in the western Dakotas and Nebraska, in token groups in far-western Kansas and Oklahoma, abundantly in the Trans-Pecos region of Texas. Yet within the vast expanse of range from north to south it has pushed into several drastically differing areas, from the aspen and conifers of the Rockies to the rain forests of the Pacific slope to the arid deserts of the southwestern United States and central Mexico.

Although the big push of settlement and industry was from the east toward the west in early days, and the buildup of human population greater east of the Mississippi, the mule deer herds nonetheless were decimated in their time by market and hide hunters, just as the whitetails were. They were tremendously abundant, were not very wary, and lived in more open country than the whitetail, and thus were somewhat easier for market hunters to collect.

There are old records, for example, telling of a single shooter killing more than a hundred mule deer from a single stand in a mountain pass in the middle Rockies in the late 1800s. At a dollar a hide, some 1500 deer were taken in a few weeks from one area of Montana by three hide hunters, the carcasses left to rot. Tens of thousands were killed by professional hunters as food for army camps, mining settlements, and railroad crew camps, and by settlers who lived year-round on venison.

However, there were generous reservoirs of mule deer herds in remote wilderness areas of the mountains, and in the dense rain forests of the Pacific coast. By the time general settlement had been effected, hunting regulations also had been brought to bear, and during this century deer management moved to a high plane of accomplishment. The handsome mule deer was transformed from a frontier commodity to a true game species, and from a low of possibly less than 250,000 animals it was brought back to an abundance that must in numerous locations be cropped severely nowadays to keep the herd tailored to its possible range.

Although mule deer certainly are in no danger, problems do beset them that do not pertain in such severity to the whitetail. Predation is undoubtedly a serious problem in some places. Coyotes, abundant throughout most mule deer range, unquestionably account for many fawns. In the north, where deep snow with a crust may occur in winter, and where coyotes are much larger than their southern counterparts, they are at times responsible for severe predation on adult deer, which break through and founder while the coyotes do not. The mountain lion, under substantial protection, presently accounts for a good many mule deer. Wolves, bears, bobcats, and other predators collect their share. The mule deer range blankets the most populous modern ranges of

## THE MULE DEER

**Color:** Gray-brown, winter, reddish-brown, summer; differs widely among subspecies; desert mule deer pale, Rocky Mountain mule deer darker, others variable in shading; nearly black atop head; nose and muzzle band black; face and eye area gray-white; gray-white throat patch; brisket blackish; ears dark-rimmed with whitish interior; belly and inside of legs whitish; very distinctive large white rump patch encircles base of tail; tail narrow, small, white with black tip, in some subspecies broader with grizzled to blackish upper surface in varying amounts and shadings.

**Measurements, mature bucks:** To 3½ feet at shoulder; overall length 6 to 6½ feet, variable among subspecies, with Columbian blacktails and Sitka deer of Pacific coast averaging substantially smaller; ears averaging 11 inches long as compared to 7 for whitetail; tail 7 inches long as compared to 11 for whitetail; metatarsal gland on outside of hind leg as much as 5 inches long, differing among subspecies, as compared to less than 2 inches for whitetail; Pacific coast subspecies noted above smaller in ear, tail, and metatarsal measurements.

**Weight, mature bucks:** Type species (Rocky Mountain) average, 150 to 220 pounds, with occasional specimens 300 to well over 400 pounds; desert subspecies and some others smaller; Pacific coast subspecies seldom at maximum above 150 to 175.

**Antlers:** Typically with antler tines branching, as opposed to the unbranched tines of the whitetail; seldom curving around and forward over brow as in whitetail; younger bucks commonly without brow tines, and with nonbranching tines rather similar to whitetail; nontypical antlers with many points not uncommon.

**Does:** Seldom more than 150 pounds; less blocky build.

**General attributes:** Exceedingly blocky, powerful appearance; far more placid personality than the whitetail; congenially gregarious among its own kind; branched antler tines; stringlike white tail with black tip (except Pacific coast subspecies noted above, which sometimes raise tail to horizontal or even perpendicular, but without side-to-side waving as in whitetail); distantly evident white rump patch (except Pacific coast subspecies); bounding gait when running that strikes all four feet on ground at once, hind feet striking behind forefeet.

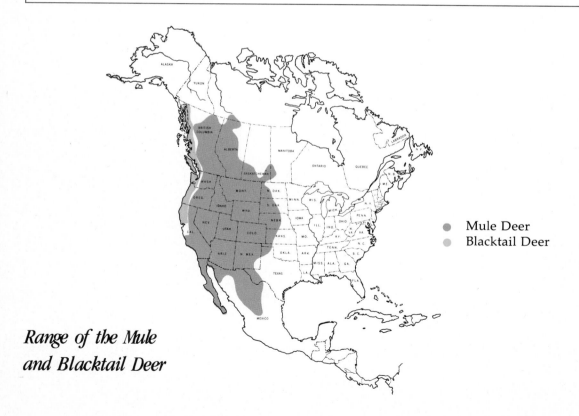

Mule Deer
Blacktail Deer

*Range of the Mule
and Blacktail Deer*

all the larger North American predators.

Starvation, and diseases often related to it, account for many mule deer. Their winter forage problem—not without exception but often—is quite different from that facing whitetails. Vast numbers of mule deer must utilize different summer and winter ranges. The quality of the winter range dictates entirely the size of any given herd. Lush summer forage may serve scores more animals than a meager or overbrowsed winter range can. Thus, starvation and disease aggravated by low nutrition unduly thin out many a mule deer herd in winter.

Perhaps, however, the most serious present difficulty facing the mule deer, the one with most influence upon the future, is the ever increasing settlement of wild portions of the west. Unlike the whitetail, the mule deer seems unable to cope with human disturbance. No one knows precisely why. Possibly because it is truly a naive product of the wilderness, it elects to retire rather than scheme toward survival on the fringes of civilization. In some instances a ski or winter resort has usurped ancestral winter range, or cut off passage to it. An entire herd that summers over a huge expanse of high country is thus doomed.

Nevertheless, at this time and for the foreseeable future the mule deer, though facing problems, is certainly not in serious trouble. Biologists dedicated to deer management in every western state diligently seek answers to some of the management questions presented by this burly and exceedingly popular citizen of the slopes and the foothills. Unquestionably they will eventually find most of them.

The Rocky Mountain mule deer receives

*Mule deer are readily distinguished from whitetails by their black-tipped, ropelike tail set in a light rump patch. Also, a buck's antlers branch like two Ys on each side, whereas the whitetail's tines rise from a single beam.*

*This is an especially fine trophy mule buck. Deep in the body, even blocky, it has massive, branching antler tines. Its neck is swollen and its eyes are red-rimmed with the breeding season. Other bucks give this one a wide berth.*

*The typically dark tail of the blacktail differs considerably from the black-tipped tail of the mule deer.*

the most attention because it is most numerous and occupies the greatest range. This "type" species is found over about 75 percent of the entire range, but southward, and along the Pacific where climate and terrain drastically differ, the mule deer has evolved into a number of other subspecies. There are ten of these. The preponderance of them fill rather small or restricted ranges and are not really very important for hunters or others to be able to recognize.

There are, for instance, the California mule deer, *Odocoileus hemionus californicus,* of some middle and southern California counties; the Inyo mule deer, *O. h. inyoensis,* of a small range along the east slope of the southern Sierra; the southern mule deer, *O. h. fulginatus,* which ranges in extreme southern California and on down into Lower California; the Peninsula mule deer, *O. h. peninsulae,* still farther south; and two subspecies named for and isolated on Tiburon and Cedros Islands. There is also the so-called burro deer, *O. h.*

*eremicus,* found along the lower Colorado River in both California and Arizona and on down the east side of the Gulf of California into Mexico.

Much more important than these is the desert mule deer, *O. h. crooki,* a handsome, pale-gray animal of the desert mountains of southeastern Arizona, southern New Mexico, and the Trans-Pecos region of western Texas, plus a range deep into central Mexico. Although this deer is smaller than the Rocky Mountain deer, in both maximum body and antler size, it is very popular with hunters, and offers a unique experience because of the unusual desert terrain.

The mule deer subspecies that is the most important, however, in addition to the Rocky Mountain deer, is one that has caused all kinds of puzzles and arguments not only among hunters but also among scientists. This is the deer of the west slope along the Pacific from portions of British Columbia to central California. It is commonly called the blacktail deer, even though some references call all mule deer blacktails. But anyone in its ranges knows what is meant by blacktail as opposed to mule deer. Originally believed to be a distinct species, and named *O. columbianus,* it was later determined to be a race of mule deer differing markedly, at least superficially, in size, color, tail, and antler development.

Some writers attempt to make no distinction between this race and the type species, a patently ridiculous stance. Properly it is the Columbian blacktail, a mule deer subspecies, *O. h. columbianus.* The Boone & Crockett records list it this way, but do not distinguish record-wise between it and the Sitka deer or Sitka blacktail, *O. h. sitkensis,* a race that is much like *columbianus,* but which extends the range northward along northern coastal British Columbia and southeastern coastal Alaska.

Interestingly enough, locations where records of these two deer have been taken indicate that no true Sitka deer yet has made the book. But the Columbian blacktail is an exceedingly abundant and important animal along the Pacific from the crest of the Cascades in Washington and Oregon to the coast and on the various islands, and on into California down to Monterey County. It is, in fact, the only deer on this slope over much of its range, although where contact is made with Rocky Mountain and California mule deer there are intergrades.

One of the most interesting theories concerning the Columbian blacktail put forth by several scientists renowned for their deer studies is that here is a perfect example of evolution at work, of a geographic race of deer drawing perhaps closer and closer to becoming a full-fledged species. It is such a different deer—the most "different" of all mule deer races—that though covered here with the mule deer it requires at least some individual attention.

Many hunters might at a cursory glance confuse the blacktail with the whitetail. It is more generally comparable in size, and its ears are shorter than the Rocky Mountain deer and are not tipped and edged with dark hair. There is no large, distinctive white rump patch. The base of the tail is brown, but the tail is always black toward the tip, and usually black on top throughout its length. It is quite brushy, reminiscent in that respect of the whitetail, but shorter. It is white beneath. Its summer coat is much redder than that of other mule deer. The antlers are seldom as broad and heavy as those of the Rocky Mountain deer, are inclined toward less spreading, more upright growth, and usually in adults they have fewer points.

## Habitat

The Columbian blacktail's world is one of dense, humid forest, or in its southern ranges thickets of heavy brush and trees often in arid settings, such as in California. It is a deer invariably of thick cover. Because of the dense

cover this deer has developed basic habits more like the whitetail's than like those of the other mule deer. When it hears suspicious sounds or scents danger, it is inclined to sneak away along trails it knows well in the thick habitat, or to stay utterly still hiding in a thicket until danger has passed. It is known to flush wildly, run a short distance, and hide again, even lying down to escape detection.

Both the Columbian blacktail and its relative in southeastern Alaska, the Sitka deer, live in heavy cover; the Sitka deer is confined to all but impenetrable spruce in a narrow strip near the coast over much of its range. Oddly, perhaps because native hunters and visitors both prefer larger game, the Sitka deer receives very little hunting pressure. It overpopulates in good years, and suffers drastic die-offs during severe winters. Sometimes it is forced to scrounge for food such as kelp right on the beaches.

This habitat is highly specialized, as are the deer. The Rocky Mountain mule deer lives in an entirely different world. Basically it is not by preference an animal of dense mountain forests. It may bed down in heavy conifer thickets, but most of its life is spent along the fringes of mountain meadows, in the mixed forest of aspen and evergreens. It is primarily a deer of open forests and of brushlands broken by openings. Typical of mule deer terrain are the mountain foothills where sage and scattered juniper merge higher up with piñon or other pines and spruces, and aspen.

In Wyoming, for example, mule deer are abundant in all such foothills lands of the Rockies. In addition they are found during summer way up at timberline, among the highest aspen growth and the fringes where trees begin to give way to low growth such as willow. Yet to the east where tree growth becomes less and the country changes to a rough and somewhat barren interspersion of grasslands and sharply eroded gullies rimmed by brush, and with steep shale hills thrusting up here and there, the deer are right at home, utilizing the meandering, deep gullies and hiding in them and in brush, far from the nearest tall trees.

The desert mule deer has colonized a still different habitat. It may be jumped from a bed right on the desert floor in Arizona, where cactus and thornbrush spread across a narrow valley with steep, rocky slopes nearby. In western Texas, in the desert mountains of the Big Bend Country, desert mule deer feed and even bed down on wholly open slopes. Though they also are found up higher here, where tree growth begins, they are always more abundant among the rimrocks and the sotol and scattered brush and shrubs of the steep lower country.

Without question the most interesting aspect of mule deer habitat is that regardless of subspecies, these deer apparently are unable to colonize level country. Why this is so no one is certain. But the history of the species shows that mule deer shun level woodlands and forests, and avoid flat or even gently rolling open grasslands. In any fringe prairie habitat where they are found, there is always a skein of rough creek or river bottoms with brush and eroded places. In western Texas, for example, deer may be observed crossing broad flats, but they are invariably moving from one slope to another. Much of the time that movement is along steep washes rather than up on the flats. And almost without fail, groups of deer observed feeding are on or at the base of a slope.

Undoubtedly this overwhelming affinity for steeply angled terrain has been an ancestral barrier to the eastward incursion of mule deer into the plains and across the Mississippi. In addition, the northward distribution is stopped by dense boreal forests of spruce. And in the south, areas of barren desert also have been a barrier, yet almost anywhere southward that ample forage on steep slopes is available, the deer have pushed in and lived successfully.

Conversely, the whitetail, which easily tolerates both mountains and flat country as long as there is ample cover, was able to extend its range almost completely from coast to coast.

## Feeding

Just as habitats of mule deer differ widely over their vast range, so too do the items of their diet. A blacktail hunter in California might watch for a buck beneath a big oak laden with and dropping acorns. In the Rocky Mountain states, however, where oaks are by no means abundant, aspen shoots or mountain mahogany might serve as the staple of diet. In the range of the desert mule deer, a slope covered with the low-growing, sharp-spiked lechuguilla leaves may be a favorite foraging ground. The deer paw out the roots and eat them.

Especially in the mountains, grasses are an important part of mule deer diet, particularly in spring. Some scientists believe that mule deer graze more persistently than whitetails. Grasses over the broad range are of infinite variety. Grama grass, fescue, bluegrass, and needle grass are among those commonly abundant. Others are brome or "cheat" grass and wheat grass. Groups of mule deer are seen all summer and throughout early fall grazing at the edges of mountain meadows. But it is in spring that grasses are the most important.

Of course during summer, the easy time of year, there are endless varieties of twigs, berries, flowers, and mushrooms to be eaten. Even in desert terrain, fruits such as those of prickly pear cactus are eagerly eaten, and also the juicy pear pads, regardless of spines, form part of the diet. Wherever oaks grow, the leaves, the acorns, and the twigs are all staples of diet. Several varieties of small oaks live in canyons and near water sources over much mule deer range. Not all oaks bear acorns every year. During good acorn years, these of course are of prime importance.

The same is true of piñon nuts. These small pines, of which there are four species, grow only in the semi-arid regions of the west. They are abundant, for example, in portions of New Mexico and southern Utah and even in high, isolated locations of western Texas. The seeds are literally small nuts, very rich and nutritious. In years when the pine nuts are a bonanza crop, mule deer gorge on them and often become unbelievably fat. Juniper berries also add at times to the diet. Mountain mahogany, a shrub with a feathery bloom, is a favorite mule deer food for browse. So is manzanita.

Sage forms a considerable part of the diet. On some winter ranges, it and bitter-brush are staples simply because the deer are forced to eat them. These are not especially nutritious or palatable. All of the fruit plants, vines, and shrubs such as grape, elderberry, raspberry, and chokecherry add variety to the diet, both with leaves and fruit in season. Summer and winter, if it is available, aspen makes up a substantial portion of the daily intake.

On winter ranges, mule deer are not inclined to dig for grasses, as do elk. Now most of their food is browse. Willow, sagebrush, and cliffrose are among the shrubs available. They also turn now to juniper and cedar, to jack pine, and on the far-northern ranges to fir. These are not especially desirable, but the deer are able, with whatever additions they can scrounge, to get through the winter on them.

It must be pointed out that while deep snow and severe cold are difficult for mule deer to endure, and cause many deaths from low nutrition and exposure, even in the warmer parts of their range the winter or non-growing season is a lean time, too. Grasses dry down, there are no fruits, no green leaves. Browse can become quite scarce, and of course the fatty crops such as piñon nuts have by late winter been consumed.

Like the whitetail deer, mule deer feed much at night, and may be especially active during moonlit nights. The routine is to feed until fairly full, then bed down for an interval, then feed again. Dawn and the first hour or so after are a period usually of heavy feeding activity. Hunters take advantage of this habit. In summer, of course, the deer feed little after the sun is well up because they dislike heat. In fall and winter, especially on cool mornings, mule deer will be out all over the slopes until possibly 9:00 a.m. Then most of them disappear, bedding down until late afternoon. By about 4:00 p.m. they begin to reappear, and feed heavily until and after dusk.

A big mule deer buck requires as much as 10 pounds of food to fill its paunch. With the variety and abundance available during the growing season this seldom takes more than an hour or two. But in winter deer are sometimes seen wandering and picking away at browse over many hours. When full they then bed down and, like all deer, spend several hours chewing small cuds which are then passed on into the second part of the stomach.

All young deer, and especially those of the mule deer ranging the Rockies where winters are severe, live a precarious existence in relation to food. Winter starvation is an ever-present threat, especially because though a summer range may be large, deer from it may be forced into a much smaller winter range. Young deer, perhaps less vigorous or unable to forge through deep snow or to reach limbs of browse trees and shrubs, suffer most. One Oregon study showed that in a four-year period some 1800 mule deer starved on a single winter range. Of these, in different years from 60 to 90 percent were fawns of the year or yearlings.

Hardships and deaths related to food do not necessarily end with the lush explosion of spring. Once green grasses blanket the range, the deer, many of them thin from malnutrition, gorge on the succulent crop. The new grass is rich in nutritive value, but it is also extremely high in water content. The quick change of diet may trigger another survival problem.

Many deer contract severe diarrhea, called "the scours." On certain ranges deaths from the scours run high.

## Movements

Mule deer are just as powerful as they look. They seem to be physically fashioned for the terrain in which they live. A startled buck goes bounding effortlessly up a slope so rough and steep that a following hunter could negotiate it only with plodding gait, pausing to rest and catch his breath every few steps. When running all out, although the mule deer is graceful, it gives the impression not of deft agility, like the whitetail, but rather of sheer power.

As has been explained, mule deer bound and land on all fours, hind feet properly behind the forefeet. Whitetails land on the forefeet and the hind feet pass them and strike next. The mule deer lands and pushes with all fours to bound again. A big buck can cover as much as 20 feet to the bound, even going up a modest grade. On the level it has been measured at 26 feet per bound. At top speed it matches the whitetail, at about 35 miles per hour. But it cannot sustain that speed for very long without panting heavily.

Sometimes the vertical height reached as a mule deer bounds is as much as 4 feet. Trapped mule deer have cleared an 8-foot fence with only a short run. The blocky, muscular build of this deer and its bounding run with the push of all fours sending it into the next leap are both unquestionably specialized adaptations to the terrain in which it lives. Even though capable of speed and long leaps, the placid mule deer when undisturbed seldom runs at all. It walks casually, big ears flopping, and is not inclined to the incessant nervous, quick motions of the whitetail. In much of its domain it can see over a large expanse, and it is calm and unconcerned.

Often when alerted but not unduly disturbed, mule deer go through an antic routine that is comical to observe. The deer stares, let's say, at a photographer who has stalked close. It looks literally amazed, yet puzzled. Almost as if embarrassed, it begins to walk away, stiffly, lifting each forefoot high, meanwhile moving its head out and back in a line parallel to the body with each step. After a few such steps, it may trot a few feet, then it begins a bounce— not a bound. It jumps up and only slightly ahead, using all four legs as springs, and coming down stifflegged only to bounce up again.

These bounces are in rather quick rhythm, but take the deer only a short distance, possibly 3 or 4 feet to each one. After this it may stop and stare back. If it finally decides things look serious, it really runs. But it may run only to a nearby ridgetop and pause again for another backward look. From there on it may simply trot over the ridge out of sight, and start right in grazing again. Out of sight, out of mind.

The majority of mule deer have little need to swim, and on some ranges probably none ever have. Nonetheless, they are powerful swimmers when need be. The Sitka blacktail of the southeastern Alaskan coast seems to have no hesitation about striking out for some distant island. They have been observed as much as 5 miles offshore, unconcernedly headed for an island at least that much farther away.

Daily movements of mule deer may extend over a somewhat wider range than in the case of the whitetail. Perhaps this is simply because their mountain habitat is so vast that it seems to beckon the animals into exploration. However, when forage is adequate and water nearby, the bailiwick in which any individual deer lives is not large. Notoriously, a trophy buck that has been reared around a certain mountain meadow will be seen there week after week. However, daily movements to food, water, and cool bedding sites may require more travel than the snug cover of the white-

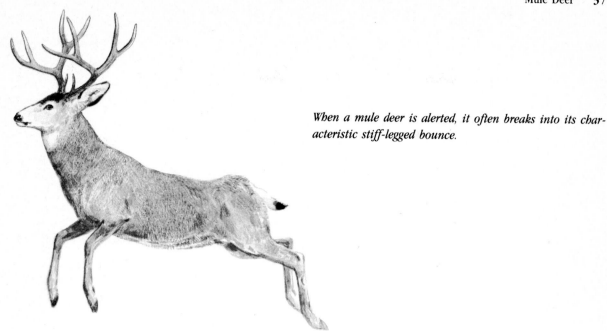

*When a mule deer is alerted, it often breaks into its characteristic stiff-legged bounce.*

tail. A deer feeding before and through dawn may have to walk a mile to water, then climb another mile up to a rimrock where it will bed down in the shade of overhanging rocks.

It should be noted that mule deer are exceedingly gregarious. Where the country is fairly open, observers may see a group of does, fawns, and a scattering of young bucks totaling as many as forty head. They may be scattered out on a slope, feeding, or even traveling in single file across a flat or up a mountain. Even mature bucks like to hang out together, something few whitetails do. On large ranches in the west where mule deer are abundant, landowners have often reported a half-dozen to a dozen mature bucks staying together during summer and fall right up until the rut begins.

The movements for which mule deer—some of them—are famous are their seasonal migrations. These are rather similar to the migrations of elk, and are caused by a need to leave the deep snows of the high country and drop down to lower elevations to a winter range. By no means do all mule deer follow this routine. The Columbian blacktails of the Pacific slope seldom need to. Some do move lower, and there are instances where mule deer move at least a short distance when they really have no need to. This is thought to be an influence from the past, perhaps based genetically.

In desert ranges there is seldom any seasonal migration. There is not enough vertical difference in altitude—often only a couple of thousand feet—to make any difference. The deer utilize the same range around the year. However, in the true high country of the Rockies most mule deer follow the seasonal migration pattern. Many of these movements are famous. In Colorado, for example, for many years hunters in several areas—Meeker is a renowned one—awaited the downward drift of the deer in order to collect a trophy. The ancestral routes are in such cases well established, and every fall as soon as snow obliterates forage up high, the deer move down to their traditional winter range.

Often a winter range may be crowded, with

herds from several sectors of nearby mountains utilizing it. In some instances late-season hunters have looked over as many as fifty trophy-size bucks on a wintering ground in a couple of days, in an area of only a few square miles. Some of the classic migrations in the west cover long distances. Instead of simply moving down to lower altitude on the same mountain, an entire herd may travel as far as 100 miles.

The fall migration may be slow, if the weather is not too severe, or it may be a forced march under pressure of high-country blizzards. The return in spring, however, is usually much slower. Now new growth moves up the slopes as snow melts and the temperature rises progressively, week by week, creeping higher and higher up the mountains. There is good forage appearing now step by step in altitude, and the deer move with its explosive growth to their final summer range. Many of the bucks arrive first. Does heavy with fawns lag somewhat behind.

## Breeding

Summer in the high country is never long. By September bucks are rubbing the dried velvet from their antlers and polishing them.

*In winter mule deer follow old trails and travel routes to reach areas, often windswept, where browse is exposed. At this time they tend also to be concentrated in large herds, which is not always in their best interests.*

*During their migrations from summer to winter areas, or when males are searching for does, mule deer do not hesitate to cross cold, fast-flowing rivers.*

The deer are now fat and sleek. A change slowly comes to them. Bucks may lie chewing a cud and watching intently a group of does and fawns nearby. The actual beginning time of the rut depends upon latitude, and to some extent upon weather. In northern latitudes breeding season is in progress about middle to late October. It may last in some instances into December. In southern portions of mule deer range the rut may not begin until November, and run on into January.

Scientists have found that the period of first severe winter storms in the mountains assists in triggering mating. Cool weather and a low barometer seem to presage mating activities. Deer that make migrations to winter range may be breeding along the way. Just as the does come into heat for only a brief period—twenty-four to thirty hours—every twenty-eight days, not all bucks are in rut at once. This is why the breeding season may extend over a fairly long period. It also virtually assures that all does will eventually be bred.

Although mule deer bucks are determined, their personality even at this important time of year is less frenzied and volatile than that

*During the rut in late autumn, bucks begin to wander in search of does. With heads uplift-
ed, they sniff the air and slash at brush in mock combat.*

of the whitetail. Their necks swell, they cease
to be as gregarious among their own sex as
previously, yet not as many serious battles ensue. On occasion several bucks may pursue a
doe and yet not enter into any violent fighting.
Usually fighting mule deer push each other
around some, but after a few charges break
off.

The does also are diligent, during their
brief periods, in pursuit of bucks that may be
haggard and therefore not especially interested. Commonly, in fact, a doe ready to be serviced tantalizes a lackadaisical buck until he ac-
quiesces. Although mule deer do not gather

harems and fight to keep them or to drive off
trespassing bucks, they do sometimes consort
with two or three does at the same time.

The bucks may travel long distances during
this period, not adhering to the comparatively
modest domain patrolled by the whitetail. A
big buck may turn up far from country it has
been living in, and even out in areas not of
suitable range. It crosses such places in search
of new conquests, perhaps on a distant mountain. This is not routine, but is fairly common.
Some scientists believe such treks are responsible for the far-flung dispersal of mule deer
and at times for the expansion of their range.

Conversely, in individual instances where deer are plentiful, a big range bull or a buck may live on and around the base of a small rock outcrop covering little more than a half-section. It will be there all summer and will not leave it during the rut. Thus is a blood line passed along to progeny at that site. Just as cattle bred to an exceptional bull drop calves to match, an especially big, vigorous, heavy-antlered mule deer buck passes along his own extraordinary qualifications. In several years another exceptional buck will replace him.

The tarsal gland on the inside of the hind legs at the hock secretes a strong, distinctive musk during the rut. Hunters often remove these with a patch of surrounding skin to avoid getting any of the wet secretion or its odor on the meat. Incidentally, a mark of distinction between mule deer and whitetail is the color difference in the hair surrounding the meta-tarsal gland, on the outside of the lower hind leg. In whitetails the hair is ordinarily white or at least mixed with white. In mule deer it is brown.

In the chapter on whitetails the hunting technique of rattling antlers to simulate a buck fight and lure bucks to it was described. The rather gentler—or at least calmer—nature of

*Whenever a doe comes into estrus, she is constantly and closely followed by a dominant buck in the area. This buck drives away aspiring bucks.*

the mule deer precludes such ready response. Experiments in rattling up mule deer have never been very successful. Some hunters believe this is because the rut falls at a different period, often outside hunting season. The fact is, most bucks just aren't that interested in getting into a fight. Does are seldom scarce and the rut is simply not that competitive. Curiously, however, mule deer will sometimes rush wildly to the wail of a hunter blowing a predator call. Possibly they mistake it for the anguished bleat of a young deer in trouble.

After breeding season is finished the bucks lose whatever belligerence they have nurtured during it and revert to their old habits. They are now thin and without great spirit. By about the middle of January in the southern part of the range, and a month earlier in the north, the antlers loosen and drop off. There is now little sex distinction or grouping. The serious business of eking out a living on the winter range or on a depleted year-round range now requires full attention. Bucks, does, and fawns now intermingle freely and with only mild bickering.

## Birth and Development

By the time spring has arrived and the movement back to the summer range has been completed, the fawns are born. Whether or not a migration is made, the fawning period is

*During the rut, a buck lowers its head and faces another buck. For several moments they clash, head to head, in a brisk shoving match that will eventually send the loser hurrying away. In such contests, serious injuries occur, but are rare.*

roughly the same, from late June on into July. The fawns are born about seven months from the time the does were bred. Interestingly, this gestation period is on the average some two weeks longer than that of the whitetail deer.

Mule deer fawns are reddish, with white-spotted coats which presumably serve as camouflage as they rest in the grass or in the dappled shade of a thicket. Does giving birth for the first time normally bear a single fawn. From then on twins are the rule and triplets not rare. The youngsters weigh 6 or 7 pounds at birth, have difficulty standing up at first, and are wobbly for a brief period. The mother coaxes the fawns to follow her as soon as they can walk, leading them to some spot she deems safe where vegetation will hide them. She leaves them, coming back every few hours to allow them to nurse.

It is an interesting commentary on planning or evolution in nature that in a few exceedingly arid locations at the far-southern end of mule deer range, the rut is much later in winter. This guarantees that the fawns will be born much later, during the one period of the year when most rains, and therefore the best opportunities for foraging, occur. It is also interesting that the milk of deer, whose young lead a most precarious and rugged existence during the nursing period and need extra sustenance, is from two to three times richer than that of domestic cattle.

The fawns stay in or very near the spot where their mother has taken them for ten days to two weeks. Then they are strong enough to follow her, and soon they are sampling the vegetation they see their mother eating. However, they continue nursing on into early fall. At that time they begin to shed their spotted coats. The new hair which comes in is their first gray winter coat. There seems to be good planning here, too. This coat is somewhat shaggy, whereas winter coats of adult mule deer, though thick, are beautifully smooth and glossy. The youngsters are not as

well prepared for winter as the adults, and thus need this little extra protection from weather.

As stated in other chapters, deer are not especially vocal creatures. However, mule deer fawns bleat like lambs when strayed from their mother. The does also "talk" occasionally, uttering a deep-pitched, coarse blatt. Mule deer also snort by blowing out air through the nose, similar to whitetails. However, although it is believed that young mule deer may bleat more than young whitetails, both adult sexes seldom are as inclined as whitetails to snort when disturbed.

By the time the winter coats of the fawns have come in, the young deer are usually weaned. They continue to follow their mother, however, and it is now that quite often large groups of does, fawns, and a scattering of young or forkhorn bucks consort. Meanwhile, the adult bucks have been off by themselves, not always seeming to avoid the others but simply spending their time as loners—particularly old bucks—or with a few of their own sex and general age group. During the rut the fawns of the year get in the way and a buck may run at them, grunting, or even give them a prod with his antlers. He is no real danger to them. The young are simply a distraction.

The fawns stay with the mother throughout the winter, and large mixed groups are common. When she is ready to give birth again, if the yearlings have not begun to wander off, she drives them away. Later, after the new fawns are tagging along, whether or not the same yearlings rejoin the same doe or not is questionable. Some probably do. Others simply tag along with the group.

During the summer the antlers of the bucks are in the thick, soft velvet stage. Young bucks are growing their first ones. Forkhorns are very common among young mule deer bucks. This means antlers with no brow tines and with a simple fork on each one. Most of these bucks run with groups of does and fawns and make up a high percentage of the annual harvest in many states. They are easier to find, with the groups of does, and quite naive.

The next year a buck may have several more points, but the antlers probably will be thin. Counting points on mule deer, incidentally, is a confusing matter to many hunters and observers of wildlife because of the traditional western method. Most hunters native to mule deer range use a term referring to only one antler, and the brow tines are not included. Thus a mule deer with brow tines and with two simply branching beams on each antler, a typical mature buck, would be called a four-pointer four on a side.

Where this can get even more confusing is with heads that have uneven numbers of points, perhaps four on one side and five on the other. Practically, therefore, it is far easier and plainer to refer to heads by counting brow tines and the total of points of both antlers, just as whitetail heads are counted. One reason brow tines are discounted in the west is that for some reason mule deer often do not have any. Further, the brow tines of mule deer are usually smaller than those of whitetails of comparable age and antler size.

In general, mule deer antlers show more variety and deviation than whitetail antlers. Some of it may be regional. For example, the first antlers of young bucks in the southern part of mule deer range show a higher percentage of spikes than farther north, where few spikes develop and most are forkhorns. Some mule deer racks just happen to form almost exactly like a set of whitetail antlers. A common formation of this sort is a head with brow tines and with main beams from which two long, unbranched points rise from either side, for a total of eight points. Such antlers when cut from the skull cannot be distinguished from similar whitetail racks.

There are also more nontypical heads among mule deer—that is, antlers with a conglomeration of points large, small, palmated, sticking out every which way. Certain geo-graphical areas produce more of these than others, although genetic influences may be responsible. As this is written, the Boone & Crockett records list almost as many nontypical as typical mule deer heads, while there are nearly a hundred more typical than nontypical whitetails.

The best mule deer antlers are grown when the bucks are six, seven, or even eight years old. Trophy mule deer antlers are always much heavier and usually rougher than those of whitetails. As mentioned earlier, they spread wide as a rule rather than curving around toward or over the brow. Some of the spreads are astonishing. Years ago one was collected from the Kaibab Plateau in northern Arizona that was 45 inches! Spreads of 30 to 35 inches are not rare. A whitetail head with a spread of 22 inches inside measurement would be an excellent one, but a similar mule deer rack would be only average. Racks of 26 to 28 inches would be far more outstanding. Although numerous points are common among mule deer, among typical heads those of ten points (eastern count—five on a side counting brow tine) are considered a kind of standard for mature bucks. The antlers of Columbian blacktails fall in general into measurement classes closer to those of the whitetail.

The quality of the winter range and of the spring growing season are directly related to antler formation. If food is scarce during the time when the antlers are forming, and the spring and summer are dry and forage is thus at a minimum, mule deer antlers will almost certainly be thin and pale in color. This is especially true in the desert ranges where severe drought is more common than in the mountains. With ample winter forage, and a lush spring, antlers will be heavy, dark, and rough at the bases, an indication of the vigor of the animals.

Color variations among mule deer are rare. Among whitetails, albinism and partial albinism are not at all rare; pure-white non-albino

*The antlers of mule deer usually have a greater spread than those of whitetails. Typically, the main beam of each antler forks part way up, and each fork has two tines.*

whitetail sports sometimes occur, and dark or melanistic specimens turn up here and there. There are shading differences from dark to light among the geographic races of mule deer, but these are natural adaptations, probably to the general shading of the habitat each occupies.

## Senses

Like all deer, mule deer have an extremely acute sense of smell. They use it to great advantage. But their rugged, steep-sloped habitat causes them to adjust use of all senses to it. For example, in all mountain terrain the flow of air is upward during the warmer hours of the day, and downward as the valleys cool. Obviously severe storms may inhibit these thermal currents, but on most days they occur. Mule deer almost without fail thus move up-

ward to bed after the sun is over the ridges. Like many mountain game animals, they seem not to expect danger from above. Meanwhile, as the warmed air from below rises, any disturbing scents are wafted up to them.

However, high bedding places are invariably selected where the deer can also see over a large expanse. Typical are the shady sides of rimrocks, rocky points overlooking a valley, or a wooded point where the deer can see out and down yet has two escape routes along either side of the point. Good observation placement therefore undoubtedly is a part of the bed selection. Like other deer, mule deer have keen eyesight, but they are not always certain of what they are seeing. Because they are colorblind, or nearly so, a patch of a highly contrasting shade—like the orange jacket of a hunter set against dark conifers—will instantly get their attention. The deer knows its domain intimately and may instantly realize that this

*Unusual mule deer bucks like this one sport nontypical, or asymmetrical, antlers. This tendency toward nontypical racks may be more common in mule deer than in whitetails. Some geographical regions produce more nontypicals than others.*

blob is out of place. But if the object doesn't move, the deer is not at all certain what it is. The animal is simply alerted, and now zeroes in its nose, and its ears, to attempt identification.

Any movement is immediately detected by the eyes of mule deer. Because of their mountain habitat, they can watch distant as well as close movement. It may not disturb them unless a scent reaches them. It is indeed common to glass an open slope and spot a big buck lying in the shadow of a single small bush. A passing vehicle on a trail down in the valley below, or a walking hunter, will not necessarily

flush this deer. It assumes it has not been seen. But it watches intently, seeing every movement and straining for scent. Thus in the fairly open habitat where most mule deer live, nose and eyes are used together constantly. In the dense habitat of the blacktail along the Pacific slope, of course, the senses are used somewhat differently. Scent obviously is as important as ever, but hearing may be more important than sight on numerous occasions. The deer hears a disturbing sound and slips quietly away, circling to pick up scent.

As an aside, mule deer fleeing at a disturbance, particularly a serious one, even such as being shot at at close range by a hunter, sometimes exhibit a most curious trait. Let's say a hunter jumps a buck at 50 yards on a slope that is dotted only with scattered shrubs and a few small thickets maybe 10 yards across surrounded by yellow grass. He shoots and misses the deer. It bounds out of sight behind one of the small thickets. He assumes in excitement that it has gone on, and is at a loss when it fails to come into view again. What it actually does is whirl and plunge into the thicket. There it stands, absolutely immobile, watching the hunter. Its combined senses seem to tell it there is no escape across the open. Some bucks get away with the ruse. Many don't.

Without question scenting ability is the most acute of mule deer senses. Hearing is keen. Some observers like to believe the big ears are a development toward uncanny hearing ability. That is doubtful. Mule deer give no evidence that they hear more acutely than the smaller-eared whitetails. It may be that they pick up more distant sounds, but this is simply because sounds normally carry farther in mountain terrain.

## Sign

Earlier in this chapter the track configuration of running mule deer was explained.

The hind feet strike behind the front feet. In some soils it might be possible to distinguish mule deer tracks from whitetail deer tracks by noting this. Running whitetails strike first on the front feet, then bring the hind feet past. But in most materials, track identification by this method would be difficult.

Of course, over much of mule deer range there are no whitetails. Yet many states have both on certain ranges, even though the whitetails may stick closer to dense cover. Within blacktail range on the Pacific slope there cannot be confusion because there are no whitetails, except a very few of the endangered Columbian whitetails in two or three locations.

Most hunters firmly believe that mule deer tracks are much larger than whitetail tracks. This is an illusion. It is true that most of the largest mule deer leave tracks slightly larger than most of the largest whitetails. The size runs roughly $3\frac{1}{4}$ inches in length for the mule deer to $2\frac{7}{8}$ inches for the whitetail. However, mule deer slightly smaller match the whitetails. Columbian blacktails, although usually smaller than Rocky Mountain mule deer, leave imprints roughly the same. Again, geographical location would identify the blacktails, except on the crest of the Cascades or at other places where the two mule deer intermingle.

The plain fact is, however, that no reputable student of animal tracks claims to be able to positively differentiate among tracks of the three deer. The same is true of trying to infallibly tell a buck track from a doe track. Identification is not very important anyway. Within mule deer range the tracks of no other big-game animals would very likely be confused with deer tracks. Conceivably pronghorn tracks might cause difficulty, where both animals use the same range. Antelope, however, have no dewclaws—which show on deer tracks in a deep, soft material—and the rear of the antelope track is broader. Mule deer tracks do differ distinctively according to the type of range. On rocky and hard ground the toes are usually worn off to some extent, and therefore leave a blunt imprint. On soft soils they are more pointed.

Mule deer droppings, like tracks, are not easily distinguished from those of other deer. But again, this is not important. On summer foods that are soft and succulent, the droppings are a mass. In fall and winter when the deer eat browse and dry forage, pellets are formed, of varying shapes depending on forage type. They are from ¾ to ⅞ inch in length.

Like other deer, mule deer when rubbing velvet from the antlers break branches on shrubs and scar the bark or peel it off from a space of sapling trunk. Because of the rather open habitat of much of the range, and the broader individual domain of many bucks, these rubs are neither so abundant in any one area nor show up so prominently as those of the whitetail. The same is true of scrapes.

In some places mule deer leave browse lines on trees or shrubs. These occur as a rule only where they are hard put for forage, or where they have overpopulated. Because of careful modern management, this sign is seldom as prevalent as it once was. Other feeding signs may be spotted, however, that tell of a prosperous deer population. In western Texas, for example, slopes covered with lechuguilla, a plant that grows densely with broad, spike-ended leaves about a foot high, are a favorite feeding ground. The deer paw out or chew out the base of the plants. Javelina also do this, but they rip out the leaves and scatter them.

A caution about mule deer sign concerns the discovery of numerous shed antlers. By the time they are seen—chiefly in fall by hunters—they will be whitened by weathering. These are antlers dropped the previous winter. And, in any area where mule deer make vertical seasonal migrations, most of the antlers will be on the winter range. Thus an abundance of them would indicate a place on which hunting time should not be wasted, unless it

is very late in the season. Oddly, droppings may lie on winter range for the winter, summer, and into the fall, and if they happen to be wet from rain or dew, at a cursory glance they can be confused with fresh ones. Thus, careful appraisal on winter ranges is necessary.

## Hunting

Mule deer hunting has evolved numerous practices adapted to the specialized habitat and habits of the animals. The technique of locating a migration route and taking a stand in a narrow pass or other likely spot at proper season has already been mentioned. Knowing the location of a winter range and hunting it—given open season—after the deer have arrived for the winter is at times almost too easy. Deer are concentrated on relatively small acreages. A few states offer early-season high-country trophy hunting. This, of course, is on the summer range. Sometimes the deer are still in velvet. The hunt is challenging, because it is certain to be rugged, at high altitude, and with the deer scattered.

In general, because of mountain habitat, glassing is far more important when hunting mule deer than when hunting whitetails. On large ranches or in National Forest lands where vehicle trails cover much territory, many hunters cruise slowly along them, not so much intent on jumping deer—which quickly move back from such trails once the season opens—but pausing to carefully glass distant slopes every few minutes. A great many deer are located thus, and stalks made. The gregarious habits of mule deer are advantageous to a hunter. A whole herd of does and fawns may be spotted together. If one has an antlerless permit, taking one of the group is seldom very difficult. Young bucks may be with the does too. It is true that very old bucks are often loners, super-wise and retiring. But groups of four- and five-year-olds consort commonly. This makes spotting them easier.

Glassing open slopes with scattered cover at dawn and for an hour or so after locates feeding bucks. Experienced hunters never are eager to shoot the first deer thus spotted. Often a slow glassing of the entire slope will show four or five or even a dozen bucks scattered along it. This habit is an assist in selecting a trophy. Glassing during the day is just as important. Mule deer will commonly bed down in places that would terrify a whitetail—right out in open sage on a slope, or under a single juniper, or in the shade of a single bush or rock. If the region is arid and the weather hot, mule deer thus bedded will lick their noses to keep them moist. These shine in reflected light inside the spot of shade like a discarded bottle or light bulb. Looking for such unnatural shiny spots has pinpointed many a deer, sometimes a whole group.

Over much of mule deer range, hunting is done on horseback simply because of the vastness of the territory. It should be done slowly and quietly, however. Horses will frighten deer, in an area much hunted, and they will run ahead or duck back around. Prowling on foot after tying a horse is a better method, or else the hunter can have a vehicle drop him off and arrange for a pickup later in the day at a specified meeting place. Although glassing from valleys finds deer, whenever possible it is best to get above the deer or their presumed locations to make stalks. They are not much inclined to watch above.

For the hunter who is a crack shot on running game, prowling the shady side of the rimrocks during midday is a productive method. Fat mule deer do not like heat. A day that may seem cool to a hunter may still be too hot for a deer to be in sun. Sometimes they even get up as the sun moves and change sides of a ridge to regain shade again. When hunting the rims or in certain terrain the headers of eroded gullies, bucks may be jumped from beds. They'll duck and run swiftly, but an expert who plans his approaches can often collect.

The giveaway sign on standing deer in open country is the white rump patch. Distantly these may appear to be pale rocks on a slope. Glassing turns them into deer. The blacktail, of course, has no readily discernible rump patch. It could not be seen very far in the cover common to blacktail habitat anyway. When mule deer are startled and run off, most of the time they go up. A hunter not adept at running shots should hold fire. Even mature bucks will usually pause at ridgetop for a backward look. If the deer goes over the ridge or disappears into timber, a short wait and then a stalk with the breeze in one's face or across may put a hunter in range. Many a mule deer settles down quickly once danger is out of sight.

When hard pressed, however, mule deer will move back into remote country. Whitetails might stay on the same range simply because they are masters at keeping out of sight. Mule deer do not like disturbance and as a rule have plenty of vast territory into which to fade.

Some hunters attempt to make drives on mule deer. This is common with coastal blacktails. Hunters are stationed on a canyon rim, and drivers move along the canyon floor. This works fairly well if one knows the terrain intimately. But the deer are still quick to spot an opening and skirt the drivers or lie down and coolly let them pass. Drives in big-mountain country can sometimes be successfully arranged by placing a hunter or two atop a forested mountain where several draws run up from the bottom of the slope and top out fairly close together. A couple of hunters on horseback riding the brush and timber down low will push deer out. The deer will drift up the draws. If the wind is carefully considered, they may come right to properly stationed hunters up on top. By and large, however, the drive is not a popular method in Rocky Mountain and desert mule deer country.

Although hunting from a stand during a migration is productive, stand hunting is not nearly as popular with mule deer as in whitetail hunting. The mountain country is simply too vast and there are too many places deer may wander. Rarely a trail can be located going, perhaps, from a watering place up to a rimrock bedding ground. If it shows much use, this may be a good stand. Or there may be well-used trails leading from slope to slope. Nonetheless, cruising via horseback or vehicle and glassing incessantly, or combining quiet still hunting with pauses for distant glassing where the cover is open enough, are the basic methods of mule deer hunting.

Most hunters favor a flat-shooting but powerful rifle for mule deer hunting, and a scope either of variable power or at least of 4-power. Shots may be long in the mountains. Big bucks can be tough to put down. Although in qualified hands the .243 does well, any of the calibers comparable to the .30/06 and .270 are standard.

There will probably always be arguments between eastern and western hunters as to which venison is best. On the average a fat mule deer is invariably fatter than a fat whitetail. It is a calm animal, far less nervous. Undoubtedly this is partly responsible. As with any of the horned and antlered animals, bucks killed at the peak of the rut or immediately after are not very desirable table fare. On the whole, although both whitetail and mule deer are excellent meat, the mule deer probably has a slight edge.

As stated earlier in this chapter, the mule deer does have some problems nowadays, mostly because of the heavy thrust of human population into the wilder country of the west. But it is still abundant and certainly will be for the foreseeable future. Although the whitetail is graceful and revered, the rugged, muscular true trophy buck mule deer is without any question the most strikingly handsome of North American deer.

*A packboard or just the external frame of a backpack comes in handy for lugging out heads and capes. This is the easiest way to carry up to 50 pounds.*

# Elk

## Cervus canadensis

*Two California tule elk bulls are silhouetted against a late-afternoon sky. At one point, this rare race of elk had been almost exterminated.*

**T**he American elk has often been referred to as the monarch of the forest. No title could be more apt. The bull elk unquestionably has the most regal bearing of all North American deer, seemingly haughty, arrogant, and totally untamed. The animal is almost perfectly proportioned, with not the slightest suggestion of ill design by nature. Although elk in modern days succumb to man's blandishments occasionally in winter—the offer of food when times are difficult—in personality this is the wildest of the continent's deer, a true creature of the purest wilderness terrain still extant within its range. Among North American antlered animals it is second only to the moose in size, and not by very much at that. Of the large members of this continent's deer family it is by all odds the most handsome.

Indians called the elk *wapiti,* a name which numerous staid references even to this day insist is the only correct one. Hunters and observers of wildlife have never accepted the term. Some of the early insistence on the Indian name stemmed from the fact that originally the animal we call "moose" was called "elk" in Europe. This was a confusion to early settlers on this continent, at a time when elk were present and well known to them over much of the eastern part of the continent. To further compound the confusion, the red stag or red deer of Europe is close kin to the American elk, and rather similar in general appearance.

The elk of the present are almost entirely animals of the high mountains of the west, with only a few scattered herds, small, isolated, and only moderately successful, elsewhere that have developed from transplants within this century. Interestingly enough, large numbers of elk in the early west were creatures not of forests and mountains but of the plains along the river courses, for example in eastern Montana, Wyoming, and the Dakotas.

Because elk meat is delectable and has a flavor not at all controversial, the elk had difficult times during this continent's early settlement and onward even into this century. Elk were virtually wiped out in the east by meat hunting and human settlement. Westward, elk meat fed explorers, settlements, and whole army camps. Soon, too, elk leather was appraised as a valuable product, and hide hunters slaughtered tens of thousands. One record from Ft. Benton, Montana, notes a Missouri riverboat loading of 33,000 elk hides in a single shipment to St. Louis.

In combination with and following slaughter of elk for meat and hides was the fad for elk teeth, the two unique rudimentary canines or "tusks" of the upper jaw that match with none in the lower and thus have no known use. These, polished and worked by jewelers and others, gained amazing popularity not only with the Elks organization but with the general populace. No one knows how many elk were killed for this single pair of teeth alone.

Thus, because of all these influences, elk were brought by the turn of this century to the brink of extinction. There is no question whatever that the modern sport hunter and the game-management experts his money

hired saved this continent's elk. From a range of much of the northern half of the continental United States and small portions of lower Canada, remnant elk herds were left in the early 1900s only in the general Yellowstone region, on the Olympic Peninsula, and in the Prairie Provinces.

Today, thanks to astute and farsighted management over several decades—paid for entirely by sportsmen's money—elk are currently present in abundance, which is to say they inhabit all terrain reasonably suitable to them, in numbers carefully tailored to the range and the forage availabe. No one knows the exact number of elk presently existent, but estimates that are probably quite accurate set the total running average at upward of 500,000. Today, amazingly, hunters are allowed to take, after meticulous surveys, as many elk *annually* as were in existence early in this century— from 70,000 to 80,000 average—and herds everywhere remain surprisingly stable and vigorous. The elk story is indeed one of the several truly great wildlife conservation achievements of this century.

The races and subspecies of elk originally on the continent are conjectural, and not wholly agreed upon by taxonomists. The eastern elk, the first race seen by colonists, was named *Cervus canadensis canadensis,* and eventually the elk of the intermountain region became *C. c. nelsoni.* Generally today the elk of the Rockies and the prototype eastern elk are considered to have been the same. The original elk of the southwest—western Texas, New Mexico, Arizona—were *C. merriami,* Merriam's elk, now extinct but replaced by transplants of the Rocky Mountain type. The small Tule elk is left only as a protected remnant in the Owens Valley in California. In the southern Prairie Provinces of Canada an elk darker and with antlers smaller on the average than the Rocky Mountain variety is found, *C. C. manitobensis.* The dark, heavy Roosevelt or Olympic elk inhabits the Pacific coastal forests.

## THE ELK

**Color:** Distantly, brown fall and winter, but in summer reddish-brown; fall coat, close-range, body much lighter brown than head, neck, legs, and with sides and flanks of bulls often still more so; brisket, belly, back line, neck mane nearly black; rump and short tail pale yellow-tan.

**Measurements, mature bulls:** Differs among subspecies; overall length 7½ to 10 feet; height at withers 4½ to 5 feet.

**Weight, mature bulls:** 700 to 1000 pounds; Alaska introductions occasionally somewhat larger, Rocky Mountain elk, *C. canadensis canadensis,* is the standard; Roosevelt or Olympic subspecies, *C. c. roosevelti,* slightly larger (and darker-colored); Tule elk, *C. nannodes,* much smaller.

**Antlers:** High, heavy, branching, sweeping up, out and backward; typically in mature bulls six points to the side, occasionally more or uneven by sides, seldom nontypical or freakish; up to 5 feet or more beam length, spread 4 to 5 feet; weight 40-50 pounds.

**Cows:** Much smaller than bulls, averaging 30 percent less in weight; generally paler in color.

**General attributes:** Exceedingly blocky; powerful build; leg-length-to-body proportions well matched; regal bearing; mane on underside of neck distinctive, very pale rump area likewise; stub tail.

## *Range of the Elk*

Thus the important elk today, by numbers and expanse of range, are the Rocky Mountain and the Roosevelt. The first is reestablished over most of the mountain states, and present in small transplant and token herds in such places as the northern Lower Peninsula of Michigan, the Black Hills of South Dakota, and a few counties of Oklahoma. The second ranges coastal mountains from British Columbia to northern California. Elk therefore are by no means endangered. It is doubtful that their range can be expanded substantially, but unless human interference destroys high country and coastal forests, elk under careful management will continue to be present in abundance.

Animal enemies are few. Some calves are killed by bears, lions, and coyotes, but predation is not an important influence. Nor is disease especially important. Several states trap sample animals from individual herds annually to test them for diseases. Overpopulation is locally the worst problem, where low-altitude winter ranges cannot sustain the number of animals that thrive on high-country summer range. Some starve, or commit depredations upon ranch haystacks, or must be fed, as at the refuge at Jackson, Wyoming, each winter. Tailoring herd size by hunting quotas is to date the best tool found to keep the elk bands healthy and vigorous.

## Habitat

Elk in the present day are, as noted, almost entirely mountain animals. Both in the Rockies and along the coast ranges of the Pacific they live out their lives on the steep slopes all the way from above timberline in summer down to winter range in the valleys and foothills. Many of the Wilderness and Primitive Areas of the National Forests are among the most productive habitats for elk.

Although transplants of Rocky Mountain elk

*A bull elk is a tremendously powerful, well-coordinated animal. It can run with a bounding gallop at 35 miles per hour, trot for miles on end at 20 plus, and hurdle a ranch fence with grace and ease. Or, if pressed, it might charge right through a fence. At 700 to 1000 pounds, bulls weigh much more than cows.*

*A bull's bugle begins in a low, rasping pitch, rises as the bull stretches its neck and partly opens its mouth, and ascends across several octaves, breaking and dropping suddenly to a grating harsh scream followed by one or more rough coughs or grunts. At a distance there is a bell-like quality to the bugle, but at close range the sound is hair-raising.*

to such places as eastern Oklahoma, where the forest is heavy and the terrain hilly, have done fairly well, the animals are fundamentally attuned to a colder clime. A few large ranches in Texas have of late experimented with elk. To be sure, they sustain themselves and produce offspring, but they are never as lusty and robust as on their natural range, where summers are crisp and the seasons definite and often severe. Most typical of elk habitat is the mixed forest of the western high country. Here spruce, fir, and various pines intermingle with stands of aspen and other deciduous trees. Some of the confier areas are awesomely dense, with jackstraw debris of blowdowns covering its floor. These offer resting places,

and it is amazing how easily the huge bulls, in fuller antler, move through the thick timber.

All of the best elk range is, like most of the mountain country, interspersed liberally with mountain meadows, or, as some westerners say, parks. Fringing the meadows are the quaking aspen stands. This tree, which grows over most of the elk range, is in fact associated in the mind of every elk hunter or observer with the animal. But it is the meadows and the open grassy slopes that are the real clue to elk presence, for the big animals are primarily grazers. They are browsers only as seasonal forage conditions force them to be.

Far up atop the high mountains, where trees run out, low willows and grassy expanses,

*This dark, heavy Roosevelt, or Olympic, elk inhabits damp coastal forests of the Pacific Northwest from northern California to British Columbia, plus Afognak Island, Alaska.*

*It is early summer and new green grass covers the mountain glades and lofty valleys. Elk cows and their calves spend long hours grazing in the morning sunshine, growing sleek new coats and acquiring fat.*

many of them surprisingly boggy, replace the aspen and conifer configurations. Summer finds many elk up here, too. All of the prime elk country is cut by swift, clear rivulets and larger mountain streams. Glacial lakes are scattered in blue droplets across the high mesas and the timberline region. Actually the elk of this modern day are confined almost entirely by the barriers of crowding civilization to the wild and more remote forests of the Rockies and the coastal ranges.

## Feeding

In the mountains latent spring seems in the final moment to come with a sudden rush. Snow still lies heavily on the highest peaks and slopes, but lush grasses swiftly carpet the meadows, pushing right to the edges of the snow. Dandelions and other early flowers paint the slopes, the pale chartreuse of new aspen leaves frames the meadows, and every rivulet is bordered by varied succulent plant life such as marsh marigold and watercress. This is a time when elk herds, having moved from their lean wintering grounds at lower altitudes to pursue the upward-exploding spring line, have the easiest time of the year.

Ribs show plainly from the sparse feeding of the winter, and the animals, now shedding winter coats, look bedraggled and far from regal. But making a living is now a simple matter of gorging anywhere they happen to be. Elk are basically grazers. Throughout their range the slopes furnish a wide variety of grasses. Wheat grass, June grass, various sedges, bluegrass, needle grass, and many others grow in profusion on the spring-moist slopes and continue on through summer and into the mild early fall. Dandelions, cinquefoil, even the piquant plants of the stream courses are all eagerly eaten.

During all of the growing season elk browse hardly at all. They have no need to. In some instances they must share their summer range with domestic stock—cattle, sheep, horses—but seldom is there a shortage of food. Even for animals as large as elk, filling up at this season is not very time-consuming. The general pattern is to begin feeding well before dawn and to continue until the paunch is filled. Then they seek a comfortable, shady spot and lie down. Like domestic cattle, elk are ruminants. Much of the day in summer is spent resting and chewing a cud.

Late in the afternoon a second feeding period occurs, lasting until dusk. By and large elk are daytime feeders. If disturbed, as in hunting season, they may stay in heavy cover during the day and come out to feed only at night. They also feed occasionally on bright moonlit nights. In late summer, when grasses have matured and turned dry and yellowed, filling up is more time-consuming. Now some light browse is also taken, such as young aspen and willow, wild blackberry, and seviceberry. When wild fruit is ripe some is eaten, and also both spring and fall mushrooms, such as the morels so common in spring in western mountains, form incidentals of diet.

When the first snows begin, the animals paw and sweep away the covering with front feet to get at dead grasses. On some elk ranges they are able at moderate altitudes to stay in the same area all winter. But in most instances the bands must move, when deep snow comes, to lower altitude, to a winter range. Elk are extremely gregarious animals most of the year. Thus bands that live together on a specific range in summer move practically en masse to winter quarters. Both ranges are commonly ancestral. Generation after generation uses them, unless human settlement or some other degrading influence forces the animals to seek new feeding grounds.

The winter is a most precarious time. Winter ranges are never as large as those of summer. In summer a whole mountain range offers abundant food. In winter deep snow to some extent impedes travel, food is of course scarce, and confinement to valleys or other suitable wintering areas limits the acreage over which the bands may roam. Now they begin taking almost anything within reach—mountain maple and mahogany, berry bushes, sagebrush, the heavier twigs of reachable aspen. And at last they turn to the conifers. Fir needles, juniper, and other evergreens furnish a substantial part of the winter diet. Browse lines up as high as the animals can reach begin to show in the timber stands.

Bark also is eaten. Elk commonly strip bark from maples when available, and from large serviceberry shrubs. They also systematically gouge out hunks of aspen bark. Most elk ranges thus show a blackened scar line, from old gouging, along a stand of aspens fringing a meadow, and on close examination new scars beside the old. Willows along winter-range creek bottoms also are eaten almost to the ground.

None of this desperation diet is very nutritious, and all of it is difficult to digest. In severe winters the animals become very thin. Now disease takes its toll. Weakened animals die from pneumonia, and from infections caused by eating too much roughage. Starvation is common. It is a rigid rule of nature that the condition and size of the winter range dictates the size of any elk herd. If overpopulation degrades the forage potential of the winter range past the point where it can recoup in summer, then regardless of how many calves are born on the lush summer range or how many animals the summer forage can fatten, the herd will be tailored automatically the next winter, by disease and starvation. This is why wildlife biologists make careful population surveys and set hunting quotas for both sexes. Good management dictates that herds must be kept at a size level that can be sustained on the winter ranges without unduly harming them.

Winter elk ranges in many states are in valleys where farms and ranches are numerous. Thus the elk get into difficulty by tearing stacks of baled hay to pieces, hay meant to tide cattle over the hard winter. They smash fences and become an intolerable nuisance to the landowners. Some states attempt to feed or disperse bands that cause complaints. And in some, state laws demand that the game department reimburse landowners for damage.

*Aspens in wintering areas often show gouge marks where elk have eaten the bark.*

## Movements

Elk are tremendously powerful, well-co-ordinated animals. A big bull can walk at a pace, when hurrying, that will lose a man trying to keep up. It can run with a bounding gallop at 35 miles an hour, trot for miles on end at 20-plus, hurdle a ranch fence with grace and ease, or even, when unduly disturbed and running, barge right through it. Supposedly "gameproof" fences at 8 feet high cannot hold elk determined to escape. They go over the top. One large ranch game preserve in northern New Mexico had to build an 11-foot fence in order to keep elk on its lands.

Although on a good summer range an elk band may stay in an area of modest size for days or weeks, distances in their vast mountain domain mean little to them. A group may drift along a slope or traverse a series of ridges, covering several miles for no apparent reason. When severely disturbed, as elk often are during hunting season in areas heavily pressured, a spooked band may take off at a run and keep right on going clear out of that part of the country, moving at a swift trot as much as 10 miles. There is an inherent wildness in elk more pronounced than in other deer.

In summer, of course, there is not often much disturbance in the high and remote wilderness country. A feeding and travel bailiwick of a few square miles will contain a resident elk population. The daily routine is to feed through the cool dawn and shortly after, then retire to timber or to a knoll from which vision is possible over a wide expanse, and rising thermals bring scents that might indicate danger. Here they bed down, and chew the cud. In the alpine meadows where most elk summer, insects are not present in irritating numbers, and breezes always blow.

Trips are made to water, but these seldom entail any long travel. Elk country is laced by streams large and small, and small lakes are often numerous. Elk are strong swimmers, and seem to enjoy splashing and rolling in cold high-country waters in summer, even occasionally swimming across a lake apparently just to get to the other side. Salt licks, commonly found in moist places or beside lakes and streams, are visited often. Ordinarily these do not show visual evidence of salt, but the earth contains it and the animals chew it, or gulp soupy muck.

When the bulls are in full antler, they are capable of slipping through dense timber with uncanny ease, often seeming not to touch a branch as they move. The nose is held high and the antlers laid along the back. In fact, heavy-antlered bulls can run all-out through timber the same way, without catching a tine on a branch. When running, all elk, even the cows, hold the nose characteristically high and outstretched, as if alert to every sound and scent carried down the breeze.

Although there is some separation of the sexes during summer months, there are no special movements to keep apart. Several bulls may hang around together, cows and calves consort, but the herd instinct, the gregarious nature, is still strong. Elk in this respect are much like cattle.

It is when first snows come to the highest summer range that the truly distinctive movement of the elk bands begins. The animals start to drift downward, sensing that food in the alpine meadows will soon be covered by deep snow. The rut still takes place in the high country, but usually well below timberline, down in the aspens, conifers, and meadows at middle elevations. But finally the exodus to the winter range is launched in earnest. This is a phenomenon like none other in nature. Some animals, such as mule deer in high

*Elk herds make annual migrations, both vertically and laterally, leaving summer ranges at high altitudes to find forage on traditional wintering grounds at lower elevations. In some areas the herds must travel long distances.*

mountains, make seasonal migrations. But seldom are the movements as long as those of the longest elk treks.

As noted earlier, not all elk herds find it necessary to make the downward migration. In a few specialized instances summer and winter ranges are one. Nor do all that travel move the same average distance. For some groups the trek is only a few downhill miles. But for others the ancestral routes, followed in some instances almost exactly year after year, may be anywhere from 25 to 100 miles. These long vertical migrations are of course more common among Rocky Mountain elk than those of the Roosevelt race, which dwell in a much milder climate.

In general, when the downward drift first starts, small groups of bulls move together. Straggling along behind come the cows and calves. There may be anywhere from a dozen to fifty of those moving together. As the rut begins the bulls scatter and take over their harems, then after the rut is finished the urgency to reach the winter range sends all the animals down toward the lower valleys. Because of human intrusion, numerous established migration routes have been cut off or changed, or the elk have been forced to winter on ranch lands or in some cases—as in Jackson, Wyoming—on refuges.

On some migration routes, individual groups from several series of summer-range and breeding-range slopes and ridges join as they move. The movement is chiefly during the hours of darkness. Then, if a slope or upper valley has been reached by dawn that offers ample forage, the herd may pause to feed and rest a day or so before moving on again toward the lower valleys. On some winter ranges a number of herds are forced to join. During the peak of the movement as many as a hundred or more elk may be traveling together. Weather conditions during any particular fall dictate the speed or casualness of the movement, and how bunched or strung out the elk may be.

When spring finally touches the valleys, the reverse trip begins. This one may not be as hurried. Forage growth moves progressively up the slopes, and the animals can move with it, feeding as they go. Now, too, most of the cows are heavy with their calves. Here and there a number will drop out of the band at moderate elevations to give birth and then later on move upward to the higher altitudes the bulls and the yearlings have long since reached.

## Breeding

During August there is already a promise of coming fall in the mountain air. Bulls that have grown fat together and placidly consorted over the summer begin to be edgy as the velvet shrivels and dries on their now-hardened antlers. They rub off the velvet and polish and clean their antlers, using spruce or other conifer saplings about 2 inches in trunk diameter as favorite rubbing trees. Then after the antlers are cleaned they begin to rip at brush and trees in mock fights.

It seems fitting that the breeding season for elk is the most beautiful time of year over most of its range. Aspens draw stunning swaths of gold across the slopes by middle to late September, first snows may lightly cloak the highest peaks, and the dark conifers form a backdrop to set off the beauty of the frost-touched landscape. Necks of the bulls, now engorged with blood, are swollen. They begin gathering harems, and bugling.

The bull elk is a true sultan. No other American deer gathers at maximum so large a number of cows. A dozen, often with calves tagging along, is routine. Twenty to thirty are common, and big bulls have been observed with forty to sixty. Usually a bull simply takes over a herd of cows that has started to migrate toward the winter range. But if he can force others to join up, or take them away from another harem, he is ready and willing. The bull elk is a classic male chauvinist. He brooks no

*During the breeding season, this Rocky Mountain bull is herding its harem to the near side of the river in an effort to use the river as a barrier to rival bulls. Reluctant cows might receive a rough prod from the bull's antlers.*

nonsense from the group of cows. Typically the bull does not lead the harem. He drives them along in front, giving slow individuals the prod of an antler, and in no gentle terms.

The bull now makes a wallow by digging in soft black earth at the edge of a meadow or beside a lake. He urinates in the mud, wallows in it, plastering himself. A mature bull in full rut is a wild-appearing, smelly, raunchy creature indeed. Wild-eyed, nose running, caked sometimes with dripping muck, it urinates and ejaculates on its own hocks and belly, and screams defiance until the mountains ring.

The so-called "bugle" of the bull is a sound like none other in nature. It begins on a low, rasping pitch, rises as the animal stretches out its neck and partly opens its mouth, and ascends across several octaves, breaking then and dropping to a grating, harsh scream, all of this followed by one or more rough coughs or grunts. At a distance there is a bell-like quality, but at close range it is a hair-raising sound.

Bulls have quite individual voices. Spikes have a reedy quality, but old, mature bulls are far louder and awesomely rough. Hunters learn to distinguish the sounds of the big bulls, and upon this bugling during rut is based the technique of calling bulls by using a "whistle" made to imitate the bugle. To a listening bull the sound is both a challenge and a warning. A lone bull may come to it, imagining he is going to be able to steal a part of a harem.

The bulls eat little if at all during the rut, which lasts from four to six weeks. They patrol the harem fringes and incessantly service cows as they come in heat. The highest activity is at dawn and during late afternoon, although bugling may occasionally be heard, usually by a haremless bull, at any time of day. There are always smaller bulls hanging around the fringes of the harem. The old sultan charges them, but he tries to avoid lengthy battles because this is the perfect opportunity for yet another bull to run off his cows.

Young bulls seldom have the spirit to take on a big one in his prime. On occasion, however, two fairly evenly matched do enter all-

*A bugling sultan bull will endure a near constant struggle to keep its harem in a tight group, safe from the attentions of other bulls.*

*During the early part of the rut in some areas, elk bulls will wallow in small pools and in mud. In time these pools smell foul enough to be detected from a goodly distance.*

out battle. The scene is one of primitive fury and ruthlessness. Now and then a bull is seriously hurt, or a sultan, beginning to grow past his prime, is deposed. On rare occasions antlers become locked and the bulls die, of broken necks or simply because neither can twist free. Antlers are often broken during fights. Numerous quick-trigger hunters have been disappointed to walk up to what they thought was a stunning trophy only to find half an antler snapped off.

Most elk calls used by hunters are tuned to mimic the thin, reedy sound of a spike— that is, a youngster with its first set of non-branching antlers. Having these pipsqueaks hang around and challenge them infuriates older bulls, which love to make a run at the youths. When a mature but still young bull does gather a harem, it commonly gives up its claim in a hurry if a big fellow moves in.

When the harem-gathering process first starts on a range where there are many elks, it does not take long to establish which bulls are the conquerors. An old bull just short of over the hill may take over a herd of cows and calves but quickly be deposed, without a fight, by some vigorous six-pointer. Or he may be challenged, attempt to fight, and be swiftly

*The Tule elk survives only in California as a protected remnant in the Owens Valley and on a few California wildlife refuges.*

*When bull elk clash, the sound of rattling antlers carries far from the meadow arena. But despite the sometimes savage collisions, serious injuries to combatants are rare. Most often the obvious loser simply withdraws from contact and leaves the area.*

whipped. The old one then wanders the ridges, constantly bugling, but usually afraid from there on to tackle another challenger.

Bull elk are unpredictable during the rut, and short-tempered. They have been known to chase saddle horses and riders, to charge hunters, and even to become a nuisance around range cattle. As the rut period wanes, however, the bulls are more than docile. They are totally exhausted and bedraggled. Ribs prominent, a bull itching to fight and service cows a few weeks previously now all but staggers along, often with head low, barely able to carry its heavy antlers. Now the bulls must begin feeding heavily, gaining back fat, for winter is not far off. All told, the breeding period for elk, and the accompanying display, is one of the most rigorous in nature, a phenomenon of procreation unmatched among the larger animals.

## Birth and Development

Late in the winter the bulls lose their ant-lers, and as spring moves closer the cows show irritation with the yearling calves that still fol-low them. Some calves have already drifted apart from their mothers. In late May or early June, with the movement back to summer range well started, the first new calves are born. Individual cows drop off from the mov-ing bands. Some calves are born in dense thickets, hidden away by their mothers. Many cows, however, seek open meadow or grass-lands where there is almost no cover at all. Yearling calves move along up the slopes with the bulls, the barren cows, and the few cows not yet ready to give birth.

Twin elk calves are not common. The rule is a single offspring. The average calf weighs about 35 pounds at birth. It is dark, rich brown in color, with white spots like those that dec-orate whitetail and mule deer at birth. Pre-sumably the spots serve as a kind of camou-flage of light and shadow when the wobbly calf lies hidden in grass or brush.

Elk are more vocal than other deer. The fawns bleat in a high-pitched voice, the cows reply in deeper tone. The cows also occasion-ally bugle, much like the bulls in rut but by no means as loud. Cows banding together with calves in summer often emit a sharp bark, a warning sound. And when groups of elk of both sexes and varied ages feed or water to-gether during summer they rather commonly utter varied bleats, barks, and squeals.

It does not take a newborn calf long to learn to get around. The mother may seek a better hiding place than the spot where it was born, nurse it, and then shortly coax it to fol-low in slow stages as she seeks the company of other cows with new calves. The urge for banding together, the herd instinct, is domi-nant. And there is greater safety for the calves when a number of cows join. There is little bickering among them. In fact, if danger looms, such as a prowling coyote intent upon seizing a calf, the entire band of cows is likely to rush the intruder. The animals are excep-tionally brave and determined where the young are concerned.

For the first four or five weeks the calves nurse their mothers, but by then they are be-ginning to try nibbling at grasses and leaves. The cows push them away now, allowing only brief nursing periods. Calves protest vocally, but more and more the cows are irritable with them, and the weaning process continues sometimes well into fall. A gawky calf now must drop to its knees to nurse, and the cow is not patient for long. The white spots have disappeared at least by middle to late August. The typical pale rump patch now shows plain-ly, and in a few weeks the long calf hair will be shed and a full new coat will grow in for the coming winter.

During the same weeks of summer the bulls have hung around in small groups, or some of them have remained loners. Their new antlers have been growing swiftly, cov-ered with velvet. The long-yearling males—calves of the previous year—have been sprouting their first antlers. Except for rare freaks, these first antlers are simply spikes, from 8 to 15 inches or more long.

Unlike most other deer, the normal two-year-old bull does not go through a "forkhorn" stage. It moves from spikes to five points on a side, or, rarely, four. These antlers will, how-ever, usually be rather thin and light. In the third year, if the bull is in good health, it or-dinarily becomes a "six-pointer"—six on a side. But these antlers also are likely to be fairly slender. From here on each year the antlers are likely to remain six on a side, but grow more and more massive. However, in some instances unusual bulls show seven on each side, or six and eight, seven and eight, or even one antler with nine. The true trophy elk are bulls that have remained exceptionally vig-orous at six, seven, or eight years old. In re-cord scoring today the number of points is not

*From late spring through summer, the rapidly growing antlers of bull elk are covered with velvet. The new antlers of the year begin to grow immediately after the old ones fall off at the end of the winter.*

as important as the beam length, width of spread, tine length, and massiveness (girth) of the main beams.

## Senses

There is no question that of the several elk senses the sense of smell is the most acute. In its mountain and forest habitat there is almost always some air movement, either from whimsical breezes that seem to bend around the ridges, or from rising thermals over the warm part of the day and downdraft thermals as the valleys cool. Elk in fact quite commonly take up feeding or bedding positions in accordance with these air movements. Under ideal scenting conditions for the animals, no hunter could get close.

However, two factors related to habitat force elk to use their ears as much as their nose. One is the broken nature of the forest and meadows, and the other in conjunction is the fact that air currents are forever eddying and swirling in steep terrain. As any high-

*Though impressive by most people's standards, this rutting bull sports a comparatively light set of antlers—six to a side. This bull is probably about four years old and will grow increasingly massive antlers in coming years.*

country hunter or wildlife observer knows, a breeze may be straight into one's face for five minutes and then suddenly hit the back of the neck. Thus, although scent is dominant, it cannot always be trusted, and the elk know this. They are therefore forever listening, bringing these two main senses to bear to solve any problems of danger. Their hearing is indeed acute. Further, like the whitetail deer, any disturbing scent or sound is not likely to make an elk curious, or pause to wonder what may be wrong. It runs first and wonders later! But unlike the whitetail, if thoroughly disturbed it may not stop running for several miles.

Eyesight is also keen, even though like all deer elk are virtually colorblind. Far from handicapping them, this undoubtedly makes their visual world all the easier to interpret. Like all deer also, the eyes of an elk are not well suited to properly identifying immobile objects. The slightest movement, however, is instantly noted.

Mountainous terrain broken into timber strips, meadows on the slopes, and open stream valleys are easy to scan over long distances. Any living creature that moves in the open or along the timber edges is certain to be noticed. Any danger moving through dense timber is likely to be heard. And if both those well-developed senses fail, the sense of smell seldom will. The whimsical breeze may cover an interloper briefly, but in seconds it may not. And when it does not, an elk even several hundred yards distant is certain to be informed by its nose. This battery of well-developed senses, added to an inherent wildness to match its domain, keeps the elk well posted as to suspicious happenings within its bailiwick.

## Sign

On any game range in order to recognize sign without confusion one must know what other animals are present that might leave somewhat similar sign. In elk range there are deer, but tracks and droppings of elk are larger and not easily confused. In a few Rocky Mountain states there are also moose. Moose tracks are larger, and not as rounded. Droppings are larger. Further, moose are not herd animals and so there are not likely to be the wealth of signs that elk bands leave.

Cattle present a different problem. Commonly elk and cattle are on the same ranges, and in groups. A careful observer of tracks will quickly discover that those left by large cattle are much more blocky and generally more rounded than elk tracks. Domestic calf tracks, however, are quite similar, but sharp observation shows them a bit larger than elk calf tracks, by possibly an inch in overall length, and smaller by at least an inch than an adult elk track. Tracks of adult elk measure about 4 inches long on the average.

It may well be that both cattle and elk tracks intermingle. To sort them out, look for droppings. Certainly cow droppings are not remotely like those of elk. When eating soft summer foods elk "chips" resemble to some extent those of cattle, but they are much smaller, at maximum perhaps 6 inches in diameter. On firm foods as in fall, droppings are pellets of varied shapes, always larger than those of mule deer and in lesser quantity than those of moose, with which some confusion may be possible. Elk pellets will be from ¾ inch to about 1½ inches long.

Hunters especially should keep an eye out for dropped antlers. Many weathered antlers usually indicate that this is winter range, for the antlers are dropped along toward March. Nonetheless, if dropped antlers and fresh tracks and droppings are all present, conceivably this is a year-round range.

Bark scars on aspen were noted earlier. These easily indicate how long elk have been using a range, by comparing old and fresh scars. Rubs (where bulls fight saplings) definitely tell of use during the breeding season. These further, when fresh, establish that a bull is in the general vicinity. Several of them around a mountain meadow edge state that this is a certain bull's definite bailiwick. Elk rubs are easily distinguished from deer rubs. They are at greater height, the twigs and branches are ripped over a longer reach of the sapling trunk, and the tree is usually larger than one a deer would select. Spruce saplings 8 feet or so tall and 2 inches in diameter at base are a favorite size and variety.

Salt and mineral licks were also mentioned earlier. These may not always be in mud. In the coastal rain forests of the Pacific, elk commonly tear rotten logs apart, paw at the under portions, and chew the pulpy interior. There are also wallows used by bulls in rut. In general the location of such a wallow is along the edge of a meadow, at a place where the ground is wet, with black dirt the deep topsoil, and with few rocks to interfere. Bulls paw and dig in such wallows until often one is a dozen feet across.

*In autumn just prior to the rut, bulls rub away the velvet, polish and clean their now-hardened antlers, using spruce or other conifer saplings as rubbing trees. After their antlers are cleaned, the bulls lash out at brush and small trees in mock fights.*

Close searching during the rut will turn up spots where a bull has ripped into tall grass and turf with his antlers, and also scraped angrily with forefeet. Back in the timber it is not unusual at any season to discover a cluster of beds, spots where the grass has been matted down where a group has been lying to rest. Because they are gregarious, and extremely active, roaming animals, elk leave a welter of signs. The informed observer can make a valid judgment of how many and what the herd makeup is by carefully evaluating all these indications of their presence.

## Hunting

The classic, most dramatic and thrilling approach to elk hunting is by calling during the rut, the bugling season for the bulls. It may not be the best time to collect an elk for its meat. Right at the beginning of the rut the bulls are extremely fat, and the meat is excellent. But as the rut progresses and they lose weight the meat gets tough and eventually, at peak of the rut or immediately following it, hardly fit to eat, except possibly when made into sausage. If the season for mature bulls opens as the rut begins, however, it's a fine time to be in the mountains, the easiest and most exciting time to bag a trophy.

Elk "whistles" are available from sporting goods stores. Some old hands make their own, often from electrical conduit cut into the shape of an old-fashioned whistle. The hunt begins simply by listening at or before dawn. If no bugling is heard, it's a sure bet a move to another area should be made. If several bulls are heard, the hunter generally works on the closest one. If he gets an answer he can soon deduce if the animal is moving toward him. Some bulls bugle constantly, some come silently. Some rush in wild-eyed, some sneak in. This is what makes the endeavor highly dramatic.

If a bull answers but won't come to the whistle, often it can be stalked. The hunter moves in, keeping any breeze properly in his face or across. He refrains from too much calling. As long as the bull continues to bugle, or answer, he prowls closer. Many a trophy has been taken this way as it stepped out of a timber edge into a meadow to peer across toward where the last whistle originated.

There is nowadays much hunting for "any elk"—that is, for a cow or a young bull or a yearling calf. Most states set antlerless quotas because it is necessary to crop the herds meticulously, keeping them from overpopulation. A young elk or a fat cow is by all odds the best meat animal, and elk meat at its best rates with fine beef on the table. It is easy to take a cow during bugling season by getting a bull with a harem to answer, or simply listening for bugling bulls and making a stalk. In most instances a bull will be with cows. It's a matter of selecting the one you want.

Warm weather, or the period of exhaustion following the rut, often sends bulls into dense timber. Trying to hunt one in such cover is virtually a waste of time. A hunter cannot move quietly, and if he does kill a bull, getting it out is a terrible problem. Most elk seasons are set so that hunters have two opportunities: during the bugling season, and later on after the rut, even in December, when there is snow, the animals have moved lower down, and the bulls have gained weight again.

Some hunters carefully check signs and deduce that elk are using a certain valley or meadow. They take a high stand early and late, watching for elk to come out to feed or to pass along a trail. This is often successful, for those who are able to sit still over long periods. The majority of elk hunters operate on horseback. It is difficult to approach elk from below. The technique is to ride the ridges, watching far off below and across. Or the hunters tie the horses and sneak to a ridge top, careful not to skyline themselves, and glass the far valley and the side draws.

Elk can occasionally be driven, but this requires that hunters know the terrain in detail. One or two hunters take a stand overlooking spots where elk pushed by drivers working through timber with the wind must break into the open. Although this sometimes produces success, elk are by no means easy to drive, and are likely to know the terrain better by far than the hunters. By and large, riding the high slopes during the day under cover of rising thermal air currents and taking a stand early and late are the most productive methods.

In a late season, with much snow, tracks help locate bands, and also the elk will require longer feeding periods, sometimes finding it necessary to be out much of the day. A prime rule for success is to make every attempt to

*The best way to hunt the high country for elk is on horseback. Horses enable you to cover long distances at high altitude. Horses are especially necessary when a new snow covers the ground, allowing tracking. Also elk are not as leery of horses as they are of a man afoot.*

*Wyoming outfitter and expert elk caller Gap Puche here bugles on a crisp September morning.*
*Very often he is able to coax a trophy bull into surprisingly close range with his high notes.*
*Bugling is elk hunting at its classic, dramatic best.*

hunt where there are few hunters if at all possible. Elk will not abide much disturbance, and once they take to dense timber because of harassment, hunting is exceedingly difficult.

Elk are difficult animals to put down. They are powerful and tenacious. It pays to go well armed with a substantial caliber. Many of the magnums nowadays are favored, especially because most shots at elk, except when calling, are certain to be fairly long, up to 300 yards and occasionally farther. A good scope, and binoculars, and, if you are after a true trophy, a spotting scope are all mandatory equipment. So are an ax, bone saw, and proper knives for dressing. When a big elk hits the ground, the real work begins!

Modern hunters can feel proud of their contribution to the wildlife scene by reestablishing elk over practically all suitable range still available on the continent. It has been solely the interest and the money of the sportsman that have made elk available both for sport and for the nonhunter, the casual observer of wildlife. Happily, elk are by no means endangered in the present day, and they are not likely to be as long as their ranges can be kept from destruction by human intrusion and by industry. Elk management has bloomed into a successful and well-ordered science. The bugling of bull elk is virtually certain to ring through the high forests of the west every fall for as long as those forests exist.

*Alaskan moose, which are the biggest members of the deer family worldwide, roam over lake and mountain landscapes such as this on the Alaskan Peninsula. They share the territory with grizzly bears.*

# Moose

*Alces alces*

No one could possibly look at a moose without being impressed, first by its size and then by its generally preposterous appearance. At first glance it seems to be a patchwork of nature in a wry mood, ungainly, ill-proportioned, utterly homely of mien. Its legs seem too long, its feet too large, its shoulders far overbalancing its hams, its enormous floppy upper lip and snout ridiculous.

All of this, however, is deceptive. A more careful observation, especially of a mature bull in fall with huge antlers well polished and gleaming, produces an altogether different impression. Here is a tremendously regal creature, an awesome reservoir of muscular power. Looked at with that view, homeliness fades, and one begins to realize that here is a creature brilliantly tailored to the harsh environment of northern wilderness, bitter cold, bog and forest in which it has for a million or more years been a successful colonizer throughout most of the northern areas almost around the entire world.

The long legs are precisely adapted to wading deep snow or mud. The outsized hoofs splay wide to steady and balance the great bulk in a bog. They also assist in swimming, at which moose are masters, as they need to be in their world which in many places is as much water as land. The enormously powerful shoulders and huge barrel with ample lung room adapt the animals for tireless swimming, or running if necessary, and for plunging through deep muck and dense stands of arctic scrub. The more slender hind-quarters easily trail anywhere the fore end can go. Even that ugly snout is a valuable adaptation, in a world where so large a creature must browse much of the time on tough shrubs and branches to stoke its big furnace against winter's below-zero temperatures.

Although this animal lacks the grace of what we ordinarily call a "deer," it actually is a deer, the world's largest. Relatives of North American moose are found in both Asia and Europe. The moose of this continent easily runs away with the title of largest North Amrican game animal. It is in fact the largest antlered animal in the world, and so far as is known the largest ever to have developed on earth.

As an animal personality the moose is a bit on the stodgy side, often rather sedentary and dull-witted, presumably with an intelligence level adequate only for its simple needs. Much of the time the animal is docile, yet a cow tending her calf can be a fury against an interloper, and a bull in rut extremely ill-tempered and fearless, even to taking on a train in battle, which has happened often in the northern bush.

It is a general law of nature that the larger an animal, the greater amount of living room each individual must have, and therefore, the fewer the total number. Thus moose are never abundant in any one locality in the same sense that, for example, deer may be. No one knows precisely how many there are on the continent, but across their vast range the aggregate of estimates from various states and provinces would put their numbers at somewhere between 300,000 and 500,000.

This is by no means an endangered species

at the present time. Like all creatures, moose have their ups and downs of population. In Newfoundland, for example, to which they were introduced early in this century, an exceedingly high population evolved, then plummeted, but is recovering. Animal enemies of such a large creature are, of course, few. There are only three of consequence: man, wolves, and bears. Moose have always been a prime food source for natives of the north country, more often than not standing between them and starvation in winter. In modern times they became also an important game animal. Nowadays the total of moose taken both for food only and for food and sport is thought to be about 80,000 annually.

Thus the moose is economically an extremely important wilderness resource. Fortunately, it is carefully managed nowadays in all states and provinces, with hunting quotas set annually after population surveys. Natural attrition by wolves, which is considerable, is not an important limiting factor. Bear predation is usually successful on adults only in snow that is so deep they founder yet hard enough to allow bears, with their large feet, to move on top.

The fact is, moose have lost little range in modern times, and have even greatly expanded it in some areas. This is especially true in portions of Alaska, into which they moved late in the last century and early in this one, finding lush food in massive burns that it is thought may have accounted for the large size of the Alaskan race. They also pushed down the Rockies into ranges—Wyoming, for example—where they had not been found by early explorers and trappers. Even in northern Minnesota, and in Maine, where once a moose was a rare sight indeed, usually a straggler from Canada, they have so successfully established themselves that several hunting seasons have recently been allowed. If the moose ever becomes endangered, it will be because of the shrinking of its wilderness domain by encroaching human settlement and industry.

# THE MOOSE

**Color:** Distantly, black; close-up, dark-brown upper area, paler beneath, lower legs grayish.

**Measurements, mature bulls:** Differs among subspecies; overall length 8 to 10 feet; height at withers 5½ to 7½ feet; height to top of antlers 8 to 10 feet.

**Weight, mature bulls:** Alaskan, *A. a. gigas,* largest of the subspecies, 1400 to 1800 pounds; Canadian or eastern *A. a. americana,* and northwestern, *A. a. andersoni* (close to identical and considered so in trophy records), 1000 to 1400 pounds; Shiras or Wyoming, *A. a. shirasi,* 900 to 1200 pounds.

**Antlers:** Broadly palmated, with numerous points along outer palm edges; spread 4 to 6 feet, occasionally over 6 feet, weight 40 to 90 pounds.

**Cows:** Considerably smaller in all measurements, weight 600 to 800 pounds; after spring shedding somewhat paler brown than bulls.

**General attributes:** Long, broad, flexible down-turned snout; unusually long legs (from ground to belly on mature bull up to 3½ feet); massive forequarters with humped shoulders; comparatively slender rear quarters; pendulous dewlap; mane on shoulders of hair 6 to 10 inches long; extremely abbreviated (3 to 4 inches) tail; large hooves and dewclaws; comparatively small eyes; large ears.

## *Range of the Moose*

## Habitat

A glance at the range map shows that the habitat requirements of the moose are rather specialized, and that all their range has a unique sameness. Whitetail deer, for example, have been able to establish themselves in a great variety of climates and terrains. But moose have never been able to populate areas south of a limit of severe cold, or outside areas clothed in typical forest mixtures of the northern part of the continent—that is, conifers intermingled with deciduous trees such as aspen and birch, and those abundant, hardy browse shrubs of the north, the willows.

On the northern fringes of their range, moose do not live past the tree line. They cannot, because the foods they require are not present. Long ago the species did apparently live as far south in the United States as Pennsylvania, in mountain terrain with suitable tree cover. Over vast expanses of their range there is an immense amount of water. Some of the highest moose populations, in fact, are in the Canadian provinces and Alaska, in regions liberally sprinkled with lakes.

Here the animals are almost as much at home in the water as on land. They wade, swim, slog through swamps. They do, of course, live in mountains, too, in the northern Rockies of the United States and in western Canada and in Alaska. But this is only because they found the climate and forest cover suitable. Typically the home of the moose is in dense conifers, birches, and willows of the north, where in summer the ground is wet and muddy and in winter frozen and heaped with snow.

There are, of course, forest openings in many moose habitats. Mountain meadows and open valleys are utilized for feeding and movement. Huge willow flats of the north and even treeless mountain slopes make moose easy to spot in certain locations, particularly in the northern parts of their range. But seldom will a moose be far from a thicket of fir, spruce, or willow into which it may retire for safety or drowsy comfort. Nor will the water of a stream or lake, or the mud of swamp or bog, be far away.

## Feeding

As with all animals, gathering food is the main day-to-day concern of the moose. In normal years, during spring, summer, and early fall few food-gathering difficulties are encountered, the chore of eating is not especially time-consuming, and variety of menu, even in the sparse habitat of the north country, is ample. In winter, however, diet is likely to be spartan, travel to forage is occasionally limited because of deep snow, and the lakes where succulent greens grow in summer are sheathed with ice.

Fortunately, moose do not require any broad variety. Basically they are browsers, eating leaves and brushy twigs and branches as the season demands. The low-growing willows so abundant across much of moose domain are a kind of staff of life, a preferred food wherever found. These, plus aspen and birch and various conifers, are the staples. In fact, even during the growing season when a variety of soft plants is present woody browse is the major part of diet.

*Here a cow and her calf of the previous spring await another spring on meager rations. During the dead of winter, moose tend to concentrate in loose groups where browse is most abundant, especially on willow flats.*

Aquatics—water lilies, floating duckweed, various reeds—are a favorite side dish in summer. Floating varieties such as the duckweeds are slurped off the surface. To gather stems and even roots of lilies, the animal plunges its head under. Wading, and feeding wet in the breeze across a lake, aids comfort in summer heat and assists in driving away the hordes of biting insect pests that swarm in northern forests in summer.

During the growing season the animals also eat varied grasses and sedges. But their legs are so long they cannot reach these without kneeling. A moose thus engaged presents a laughable picture, down in the fore on its knees, moving slowly along and cutting a swath of greens. To get at high leafy branches it commonly goes in the opposite direction, rearing up on its hind legs and tearing them off 10 or 12 feet above ground. It can easily

stand naturally and reach up 8 feet or more. When desirable branches are too high to reach, a moose bends saplings down to convenient level with its snout or teeth, or it rears up and rides them to ground.

Along with the staples a variety of other browse is consumed where it happens to be present. Some of the favorites are alder, raspberry, mountain ash, ground hemlock, maple, and chokecherry. During the summer not much use is made of fir, cedar, or other conifers. Those are winter browse, not as palatable or nutritious but serving when times are lean.

In winter they also eat bark of aspen when forced to, gouging out chunks with the lower teeth, which must do the job because a moose has no upper incisor teeth. Occasionally in aspen stands where moose are plentiful, bark scars, which turn black as they heal, form a visible horizontal line at moose-head height along the edge of a mountain meadow. Bark, however, is by no means a preferred item of diet.

There is no strict feeding routine. The animals may feed at night as well as by day. Normally, when food is plentiful, as in summer and fall, there are two periods of heaviest daytime activity. These are from before dawn to a couple of hours after, and again late in the afternoon. These are the times when one wishing to observe moose ordinarily has the best opportunity. Commonly at those times the animals are in the open along a lake shore or a forest-edge feeding ground. When filled up, they retire to a thicket to drowse and chew their cud.

*The moose often plunges its head below the surface of a lake to feed on bottom-growing lily stems and roots.*

*A cow moose uses her height to browse on willow saplings.*

to spring. Winter is the time when cows are pregnant. The spring calf drop is directly related to the health of the cows during those months. A bad winter after a poor growing season invariably means a meager addition of calves to the herd. Further, it is in spring when the antlers of the bulls begin to sprout. A bad winter and a slow spring point toward poor antler growth, and trophies are likely to be few the next fall.

## Movements

Notwithstanding its size and strength, the moose is not much of a traveler. Under ideal forage conditions certain individuals may live out their entire lives—on the average ten or a dozen years, an extreme of twenty—within a bailiwick of a mere 500 to 1000 acres. An area of 5 square miles contains the casual wanderings of most. In their wilderness there is not much disturbance. A pack of wolves may jump a moose in timber, forcing it to bolt for a lake and start swimming, where the wolves will not follow. During hunting season, man's intrusions disturb a few animals. But theirs is a vast domain and most of the time a farily peaceful one where the huge beasts can afford the luxury of laziness.

The major part of the daily routine is made up of eating, making short trips to water, and lying or standing sleepily in a comfortable thicket. Sometimes a big bull will stand for hours in cover, its brain presumably in neutral, enjoying the utterly simple and pleasurable animal pastime of doing nothing. In fact, often

Unusually severe winters occasionally have a disastrous effect upon a moose population. This is especially true when a summer has been exceptionally dry or cold and plant growth thus inhibited. As winter wears on, food becomes more and more scarce and snow piles up deeper and deeper. It takes a lot of snow to founder a moose, but it does happen. As the food supply dwindles and getting to it becomes more and more difficult, fat put on earlier is burned up in the continuous battle against below-zero temperatures.

In an ever weakening condition, animals are now susceptible to numerous diseases, such as pneumonia, and to internal parasites. Starvation is a real threat, and does at times literally wipe out a local moose population. As the animals lose strength, predation also becomes a factor. They are much easier kills now for wolves. Even in a normal winter, the quality of the previous growing season is directly related to how well a herd will come through prior to the rutting season, when all moose realize instinctively that they must feed well to put on fat to be drained away during the frenetic breeding activity, a big bull may adopt an especially lush expanse of savory willow as a temporary home, and gorge there for several weeks, moving out only to take on water, which is usually nearby.

This is not to imply that moose are incapable of fast and strenuous action. Far from it. The walking gait is one of long strides that eat up distance. When urgency of some kind breaks that rhythm, the animal shifts into an easy trot. An observer watching this movement is immediately aware that even in its ungainly configuration a moose does have astonishing grace. When the trot is accelerated by either anger or suspicion of danger, speed and evident power still further dispel any idea of awkwardness. The next shift in speed, when it is needed, is to a gallop. Curiously, these large deer gallop much the same as their small relatives, the whitetail, bounding, so that the rear feet strike slightly ahead of the front ones.

Moose are startling runners when under pressure, going full out for miles without seeming to tire. Amazingly, with their great bulk, they still turn in better time under stress than smaller deer such as the whitetail. They have been clocked with fair accuracy at about

*Moose readily plunge into a swift stream and swim across.*

*Especially in summer, as the new aquatic vegetation begins to emerge, moose are likely to be found at dawn and dusk near shallow waterways.*

35 miles per hour—trotting. A moose going all out in a gallop and unimpeded by mud or snow can up that, it has been estimated, by another 10 miles per hour! Unlike the smaller deer, however, moose are not jumpers, except under pressure. They are capable of clearing hurdles possibly as high as their shoulders, but they seldom encounter a need to jump. With their long legs it is simply too easy to walk over anything that gets in the way.

Swimming, previously mentioned, is a strong point. Moose commonly wade into a lake and swim straight across, covering several miles. Water, perhaps, is a shortcut, to save walking clear around the shoreline. They make

good speed, and appear tireless, although occasionally because of misjudgment one drowns.

Whether from lack of fear and confidence in their enormous strength, or from low intelligence, a moose will plunge into a swift stream and go with the current while attempting to cross, or enter without hesitation the deepest bog. Animals have been observed down to the withers in ooze, bucking away a few feet at a time, resting periodically and finally making it across the sinkhole. In summer daily trips may be made from the forage area to a mudhole beside lake or stream which serves as both a salt lick and a wallow. Or the lick and wallow may be at separate spots and visited routinely. Wallowing in mud cakes the hide to make it impervious to black flies and other biting insects. Muddy places that contain salt are eagerly sought, and several animals may use the same one. It's not a "lick" in the strictest sense—they simply eat the muck!

Over the flat to rolling parts of moose range, most of it in the eastern half, no late-fall migration is necessary. The animals stay on their home tracts around the year, often "yarding up" in winter in heavy conifer cover, several together, and making trails through deep snow that they keep open by daily foraging trips. In mountain terrain, however, they may find it necessary to make at least short vertical migrations. As snow piles up on the higher slopes and food becomes scarce, they drift down to the valleys. Rivers and creeks with stands of willow or alder along the banks are a winter home and feeding ground. When spring greens the upper country, the animals work their way back up to spend the summer.

An intriguing aspect of moose movements is that with all their size and bulk even the largest heavy-antlered bull can slip through cover without a sound when he wants to. Natives who live in moose country know this trait well. An animal spotted distantly and stalked to close range may not crash away noisily if aware of the hunter but simply fade into timber without a sound.

Moose are not herd animals, nor even especially gregarious. Bulls may be loners for months at a time, and so may cows with their calves. But certain seasonal movements of the two sexes are rather ritually followed. Cows and calves may cling to streamside thickets while bulls will be found scattered in the hills. Normally the two sexes do not consort purposely except during breeding season. In hard winters when groups band together in a yard it is apparently not a matter of seeking company but of finding the most comfortable, well-protected place.

## Breeding

It is during the fall rut, or breeding season, that both bull and cow moose are at their most active, traveling more than at any other time of year. A bull now may roam outside his home bailiwick, seeking cows. He forgets all about feeding. When he strays into new territory he may wind up in a fight, or find one waiting in his own domain with a bull also crossing home boundaries. When a fight actually occurs, it is invariably between bulls of similar age and size. A young bull may be able to breed when it is about 1½ years old, but the young ones know they are no match for six- or seven-year-olds in their prime. One may hang around, even when a large bull is with a cow. But most know better than to mix it up with a mature male, and the larger one doesn't have to make too many crashing runs at the youngster until he takes to his heels.

Bull moose do not collect a harem, as do elk. Now and then a bull may consort with two or three cows in a group. Generally, however, the bull is content with one amorous cow at a time. He stays with her a few days, then leaves her to seek a new companion. The rut begins about when leaves turn color in fall, which is roughly middle to late September in the north. Bulls have been rubbing velvet from their antlers from late August on into September, and polishing them, ripping at trees and brush in mock battles.

The rut lasts from a month to six weeks. When it starts the bull's antlers are gleaming, he is stately and arrogant, and his temper is short. Anything from a hunter to another bull to a freight train on a backbush line that gets in his path is fair game. By the end of the rut, having serviced as many willing cows as he can find, his temper is still short but he is spent, thin and bedraggled, often carrying his head low because antler weight is almost more than he can drag around.

The cow is possibly the most aggressive of

*This bull, which has been following a cow, grunts continually to advertise its intentions.*

all deer. Most are in their third year when first bred, although they may be mature enough earlier. The cow runs in excited circles, bawling hoarsely an invitation to any listening male. In fact, the breeding season is the only time of year when moose are especially vocal. Both sexes utter grunts and coughs occasionally at any time. But during the rut the raspy wail of the cow and the deep, coarse grunting of the bull indicate the excitement of this all-important season.

These vocal antics, incidentally, are the basis for calling moose by hunters and photographers. Natives mimic the sounds using a bark horn or cupped hands. Nowadays there are even modern recordings of the sounds, which sometimes bring in a bull on the run crashing through brush. Indian hunters even learned long ago to dip a bark megaphone into water and spill it out, to imitate a cow urinating in a lake. At this season that sound, used with the mimicked wail of the cow, often proves irresistible to a rut-crazed bull.

When one bull hears the coughing grunt of another, he makes no silent sneak attack. He goes crashing through timber, furiously swinging his huge antlers. This sound also is imitated by native hunters by beating brush with a stick to bring a bull into sight. It is interesting to note that during most of the year an altercation between moose is settled by rearing and slashing out with front hooves.

*Early in the fall and the rutting season, it is not uncommon to find bull moose in small herds, or bachelor clubs. As the days pass, the bulls joust and maneuver to establish rank. The strongest, most aggressive bulls will do most of the mating later on.*

*When two mature bulls fight during the rut, their fury is startling.*

This is a typical deer maneuver, especially common with whitetail does. But during the rut the bull wastes no time on such passes. He puts his head down and charges. Although there is probably not any great amount of serious fighting, when two mature bulls do tangle all-out their fury is startling. Now and then one is killed, or antlers become locked. If that occurs, both bulls usually die, one of a broken neck, the other because his victim is his own nemesis.

Wallows, previously described, are of a different kind during the rut. Bulls make them as a kind of sex symbol. On occasion a summer wallow is used. More often, however, fresh ones are made, usually at the edge of cover. Mud is dug up by the front hooves to a depth of a few inches to a foot or more. The bull urinates in it and on his own hocks, and wallows in the reeking mud sometimes until his body is caked and dripping. Probably his odor is a kind of "call" all its own to cows in the vicinity.

The seasonal occurrence of breeding is timed precariously, as with all animals living in severe winter climates. The rut for moose winds up around the beginning of November or a bit earlier. That is when winter begins to close in. Bulls must now spend all their time diligently feeding, to regain weight lost during the rut and to put on fat to be burned as winter progresses, to hold body heat and ward off cold. Cows must fatten swiftly to feed both themselves and their burgeoning offspring.

## Birth and Development

It is roughly eight months from the time a cow is bred until her calf or calves are born. This puts calving time into May or June when spring is well launched. Young cows may drop a single calf. Twins are common among older ones. The ungainly little creatures do not have white spots, as do whitetail and mule deer, but their color is quite different from their parents. They are red-brown, and weigh up to 25 pounds.

In the several weeks previous to birth the

cow becomes quarrelsome and irritable with her offspring of the previous year, if they are still following her around. They commonly do stay with their mother all year, tagging along even during the breeding season, when they are a nuisance but are tolerated by the bulls. But now as birth time for the new crop approaches, the mother makes menacing rushes at the gawky yearling until in fright or perplexity it takes the hint and either drifts away or she sneaks away and leaves it.

At birth moose calves do not have the humped shoulders of the adults, nor the huge overhanging snout. The cow seeks a secluded thicket or an island in a lake to give birth, and she and her calf stay within a small area for several weeks. Calves utter low bleats and are able to run swiftly when only a few days old, and soon they swim with their mother, getting a free ride when they tire by laying their heads across their mother's back and letting her do the legwork. After that first several weeks close to the birthplace, calves go wherever the cow goes, staying near her from then on through summer, fall and winter, until it comes their time to be driven out into the wilderness world on their own.

There is a substantial loss of young moose to various enemies. Bears and wolves, and even the lynx and bobcat, account for some newly born youngsters. All during their first year they are vulnerable because of their modest size. However, the cow is extremely watchful and determined in her attacks on interlopers. Fishermen and others horsebacking into wilderness moose country in summer occasionally have hair-raising adventures when an irate cow takes after them. Unquestionably exposure, varied diseases, and accidents take a greater toll of young moose than do predators.

The yearling chased off by its mother to begin life all on its own is possibly now more vulnerable than at any other time to large predators, and to accidental death. It is at first gid-

## Antler Growth

*Late April*

*June*

*Early September*

*Male moose, from yearlings to old bulls, begin to grow new antlers early in spring. The antlers at first look like velvet-covered buttons.*

dy, confused, naive, bumbling. There are many instances of moose-country travelers finding loner yearlings not even wary enough to show fear.

Soon, however, the yearling begins to catch on. While it is becoming oriented and experienced in its new life, the mature bulls have been doing nothing much during the summer except eating and growing new antlers, which began as tender, velvet-covered knobs and take

full shape with summer growth, after which with blood flow ceasing the velvet-rubbing and polishing process begins once more, in preparation for another breeding season. Even the long-yearling bull generally has his first antler-growing experience. These antlers aren't much. Usually they're simply spikes a few inches long.

At two the young bull pursuing a normal antler growth pattern has flattened forks. The

*This Alaskan bull, with loose velvet hanging, has just emerged from a session of rubbing his antlers on willows.*

next year his antlers begin to show the first real palmation, but the palms are narrow, small, and with few points. Each year thereafter, if good health and vigor continue, and seasons are amenable, the antlers are larger and larger. At six or seven a bull is in his prime. He may, however, if especially vigorous, continue to produce still larger, heavier antlers for ten years or more, after which, in declining old age, the antlers retrogress. In sheer weight, the antlers of a mature bull moose are heavier than those of any other of the earth's antlered animals.

## Senses

As in all deer, the sense most highly developed in moose is that of smell. Most creatures living in an environment that contains large expanses of heavy cover have an exceedingly acute sense of smell. Creatures preyed upon and living in brush and tree cover must use this sense constantly.

The same is true of hearing. Listening for danger is ever important. The ears of a moose are keen indeed. Conversely, moose do not have especially keen eyesight. Like all deer,

*A couple of days following the shedded-velvet stage, this Alaskan bull's antlers are free of velvet—the antlers red with drying blood.*

they see motion very well, but are not equipped for, or adept at, interpreting the meaning of motionless objects. Sometimes they are not even very concerned with distant moving objects.

Part of this may be because such large animals are not especially fearful. Part may also be because of a rather modest intelligence level. The senses of all animals are attuned to what use they need to make of each. In cover a moose cannot see distantly anyway, and in the open it is chiefly concerned with its immediate surroundings. So perhaps it sees all it needs to see. Further, like all deer, moose are colorblind, or nearly so. They live in a world of gray. This is not necessarily a handicap. Conceivably surroundings all in shadings from black to white may make far less confusing the chore of sorting out objects of which to be wary.

Observers, whether simply lovers of wildlife, photographers, or hunters, must always bear in mind, however, that when one animal sense is alerted, the others are brought to bear immediately like a battery of electronic devices, all trained and tuned in to bring messages to the brain. Thus with its moderate seeing abilities, when a moose spots something to be even curious about, it begins listening harder. It may circle to get the wind on the object, a position from which it also can hear better. It is therefore equipped to take good enough care of itself—as its success in staying abundant on earth for hundreds of thousands of years well attests.

## Sign

Anyone looking for moose needs to recognize the signs they leave. Because of their size, all their signs are quite obvious. Tracks are, of course, one of the most numerous. Over much of their range no other hoofed animal is resident, so there can be no confusion. However, in some places there are deer, or elk, or caribou. The size of moose tracks easily separates them from those left by whitetail or mule deer. Caribou tracks are exceedingly distinctive, the two hoof portions spread and very rounded, with rather blunt toes. Some confusion is possible between moose and elk tracks. However, moose tracks almost without fail are much more pointed, and more narrow. They are also larger among mature animals.

The medium in which tracks are imprinted—for example, mud—often exaggerates their size and shape. But a clean moose imprint of a bull may measure, the two large hoof portions, 6 to 6½ inches in length. From dewclaw prints to front point of track may be over 10 inches. Cow tracks are smaller, by about an inch.

Tracks are not always a good measure of moose abundance. It doesn't take many moose to leave a lot of tracks in the places that print them most plainly. Droppings are a sign that give both indication of presence, and a hint of abundance. The pellets are most easily identified in fall and winter when the animals are eating little succulent, green or soft food. They are rounded or elongate, occasionally rounded with one concave end, and measure 1 to 1¾ inches in length on the average. This is larger than those of elk. In summer the droppings seldom have pellet form and may appear as a soft shapeless mass. The quantity of moose droppings at any given stopping place, such as around a bed, is invariably larger than that left by elk, simply because of the animal's size.

It is doubtful that moose beds, in either vegetation or snow, can be easily distinguished from those of elk where both are on the same range. Numerous bedding spots do give some indication of abundance. Earlier, the bark signs on aspens were noted. In winter when moose feed heavily at times on fir or other conifers, a "browse line" up as high as they can reach is easily seen in such a timber stand, with all twigs and branches cleaned off. Willow expanses show many broken bushes. On winter ranges some patches are overbrowsed until all but destroyed.

Wallows, used both in summer and during the rut, and salt licks are other easily spotted signs, although not as numerous as the others. On a winter moose range where a number of animals may be forced to spend several months, shed antlers, dropped during mid-winter and usually whitened by exposure when found, offer evidence that the animals have been there and probably will be back the next winter.

## Hunting

How moose are hunted depends on the terrain. Over the eastern half of the range, which is on the average rather flat and with dense cover, hunting from a canoe is a productive and popular method until freeze-up. Sign is checked along a lake shore or a stream, and hunters patrol quietly, the best hours early and late in the day. Glassing carefully and distantly is important. A moose across a lake may look as black as the upturned muck-filled roots of a blowdown, and be casually passed up as one. With wind right, often a moose seen distantly can be approached by canoe to within rifle range. Shots should not be taken when the animal is belly-deep in water, however. Getting it ashore for dressing is an awesome task.

In the west where terrain is more mountainous, hunting is traditionally done on horseback, with long pauses to glass surrounding country from a high point. This type of hunting may continue all day, watching for moose not only when they are out early and late but also when they may be resting in thickets during the day. A hunter on foot,

however, if he is wise, checks sign first, then takes a stand early or late in the day where he can watch a likely spot for a bull to emerge into a meadow or other opening.

In either case when a bull is spotted, if he is out of range a stalk is made on foot. Keeping the breeze in one's face or crossing is mandatory. So is moving quietly and utilizing cover whenever possible. If stalking cover is lacking, the hunter moves only when the bull is feeding or has his gaze turned away. The moment the bull looks in the hunter's direction, he must freeze until scrutiny ends.

Throughout moose range, calling works rather well during the rut, as has been mentioned. Few modestly experienced moose hunters are adept at it, and they may arouse suspicion rather than eagerness in a listening bull. The majority of moose hunters are guided, and it is best to leave calling to the guide.

Many hunters like to go after moose when snow is on the ground, if the season is still open. This still hunting—and obviously it can be done at any time of season—is a matter of heading into the breeze, moving very slowly along moose trails, old logging roads, or lakeshores, or even following a track to try to jump a bull from his bed. One should go quietly, prowling, pausing every few yards to carefully scan the area within view. Often only the glimpse of an antler of a bull bedded in a thicket is seen. This is one of the sportier hunting methods, requiring much craft and patience. Still hunting or stand hunting can be combined with horsebacking or canoeing. Locating a good stand by riding, or exploring inlets with a canoe, is productive. Or a moose spotted distantly from a canoe can be stalked by keeping the craft behind a point to get as close as possible, then putting the hunter ashore for a stalk.

Moose hunters should never go under-gunned. The great bulk of a bull dictates using a heavy-caliber rifle and large bullet—and not

*Here a hunting guide attempts to lure a bull moose into range, imitating the bugle of
another bull by blowing into a horn fashioned from birchbark.*

listening to the tales of natives who've been harvesting moose all their lives with a .30/30. They pick their shots and get close. A trophy hunter may not always be able to. In addition, any hunter should be sure that he or his guide is equipped with saws, knives, and ax, as well as rope and compact winch or other equipment that may be needed to get a moose into position for dressing, and then doing the job. It is a prodigious task.

In some places any moose is legal, or there are quotas of cows and yearlings. The meat of a young moose is excellent. So is that of a fat cow. A bull taken during rut is strong and a long way from delicious. Immediately follow-ing the rut it is all but inedible. After a few weeks, however, when fat has been put on again and it is "on the mend," it is good fare. The thousands of pounds of moose meat gath-ered annually, and the outfitting and guiding business based on moose hunting, make this animal an important resource of the north country. Management methods—lengthy flying surveys to check moose populations and the meticulous setting of quotas and seasons to crop surpluses and keep the population in balance with available food supply—give this magnificent animal every chance of a bright future, perhaps in some quarters even more abundant than ever.

*The caribou is one of the most specialized of all North American deer. They range across Arctic tundra, beyond treeline, and often in great numbers all across the lonely, brooding northern landscape.*

# Caribou

*Rangifer tarandus*

The caribou is an authentic enigma, the most specialized and puzzling in personality of all North American deer. It looks and acts at different times both magnificent and ridiculous, intelligent and stupid. On this continent attempts to domesticate it have never been successful, yet its European counterpart, the reindeer, for all practical purposes identical to the New World animal, has been a domesticated, harness-broken beast of burden since early times.

Caribou are animals of the far north, where they live across the tundra far past the tree line. They also range southward into some portions of Canadian forest. In days of early settlement in the United States they were known to range in remnant bands into New England, the northern fringes of the Great Lakes states, and into Montana, Idaho, and Washington. The great herds originally found on this continent undoubtedly numbered in millions, and stretched from the Alaska Peninsula to Newfoundland, spreading northward across the Arctic islands to land's end.

Because caribou are extremely gregarious, and some of them make long cross-country seasonal migrations, all northern peoples since ancient times have utilized caribou as a staple of diet and the hides for clothing and shelter. Many still do. The herds are by no means as large today, and in certain areas they have had drastic fluctuations. They are not, however, presently endangered. In Alaska caribou are the most plentiful big-game animals, with total population estimates of 500,000 to 600,000. They are fairly plentiful in the Yukon, the Northwest Territories, and parts of British Columbia. There are herds of moderate size stretching across the northern portions of the Prairie Provinces. Eastern Canada—Quebec, New Brunswick, Labrador, Newfoundland— contains substantial numbers. Thus there are

undoubtedly well over a million still left on the continent, and almost everywhere they seem to remain fairly stable.

The personality of this animal is anything but stable. It is a whimsical character. A big bull may be feeding along as the rut approaches, suddenly rear up with forefeet pawing the air, whirl and run off some distance, then suddenly begin to graze again. Sometimes hunters find caribou as shy and wild as whitetail deer. Sometimes also the animals seem totally addled, standing to watch a hunter make his stalk. At times, apparently out of curiosity, a group or a lone bull will walk right up to examine a hunter, or even a pack or saddle horse.

When shedding their winter coats, caribou look like a bunch of ragged tramps. Yet a mature bull in its full winter raiment and with a great curved rack of antlers is a magnificent sight, as, alerted, it trots at a swift pace across the tundra. But it can as quickly change, when resting, to a clumsy-looking, dumpy, ill-proportioned creature that seems to have no sense at all. Every time it takes a step, the tendons and bones of the ankle make a clicking sound. The feet are so big they're preposterous. No two sets of antlers on the bulls are similar, and a good many are not at all symmetrical. The poorly developed antlers of the cows look like an afterthought of the Creator.

The long migrations caribou make are as whimsical as their other traits. Year after year for some seasons thousands may move together along an identical route. Then suddenly some season they fail to appear, having for reasons of their own switched travel lanes.

*It is midsummer and the caribou bull's antlers are still covered with velvet. At this time caribou are also shedding and moving constantly to avoid the myriad insects which hover over the tundra.*

## THE CARIBOU

**Color:** Winter coat, body dark brown to gray-brown; sides of neck pale gray to white; mane on underside of neck gray to white; rump patch, underside of tail, ring around each eye and above each hoof pale gray to white; varying amounts of pale gray to white running from neck and chest mane back across shoulders and along flanks; color differs, lighter or darker, among subspecies; white ruff and neck grows more distinctive as bulls age; races in northernmost Arctic nearly white.

**Measurements, mature bulls:** Differs among subspecies; overall length 6½ to 7½ feet; height at withers 3½ to 5 feet.

**Weight, mature bulls:** 300 to 400 pounds average; exceptional specimens, certain races, to 600-plus.

**Antlers:** Main beams sweeping back, up, then forward; ends of main beams, and one or more branches from each main beam, usually palmated, and with several points; unique brow tines palmated and with several offshoot points, and set vertically; these brow tines extend over the face and are called "shovels"; commonly there is only one palmated brow tine, the other a simple spike. Curve of beams may measure 4 to more than 5 feet, with greatest width nearly comparable. Antlers seldom symmetrical, and highly varied.

**Cows:** Smaller; antlered (the only American deer with this characteristic); antlers much smaller than those of male.

**General attributes:** Clumsy-appearing build; chest and neck mane of bulls in fall striking; stubby tail; blunt muzzle, haired; horselike face; extremely large feet with hooves almost round.

## *Range of the Caribou*

That idiosyncrasy is a serious matter to native peoples of the far north who wait for the herds that some winter fail to appear. People of many an outpost settlement have starved because of it.

Notwithstanding its odd traits, the caribou is a highly specialized creature almost perfectly adapted to its severe environment. In a land where it is one of the most important forage animals for the large predators—man, wolves, bears—and where most of the year is winter and weather and general climate are seemingly unbearable, it has managed for thousands of years to thrive. Not only have caribou evidenced an amazing tenacity for survival in large numbers, they have meanwhile diversified over their vast range until they've become for scientists a taxonomic nightmare.

Seldom do two zoologists agree as to how many species and subspecies there are. Some years ago one authority listed three species and fifteen subspecies. It is easy to understand how variations in animals from location to location came about. Because of their strong herding instinct, and seasonal movements, over many years numerous individual herds built up in their own specific areas. These in some cases eventually began to show different physiological characteristics, differences in size, antler development, and color. In Alaska,

*A caribou's world is often gray and gloomy—even eerie. Caribou move erratically, sometimes it seems aimlessly, the tendons and bones of the ankles making a clicking sound. The feet are extremely big, a gift of evolution that helps them move easily over soft ground.*

for example, there are presently eleven fairly distinct herds, ranging in numbers from a few thousand to over a hundred thousand.

As the herds anciently developed clear across the top of the continent, differences in habitat and geography produced animals differing enough that scientists who studied them first set up several distinct species, and then subspecies within those ranges. In Alaska there is the Barren Ground caribou, usually given the scientific name *Rangifer tarandus*. There is a subspecies on the Alaska Peninsula known as the Grant's caribou, and so it is tagged *R. t. granti*. Another nearly white race, the northernmost form, on Ellsmere Island and named for Admiral Peary, is *R. t. pearyi*.

Some taxonomists carried the subspecies system to hair-splitting lengths, and some still do. Some claim that today there are twelve subspecies on the continent, others that there are but seven. Further, the present most accepted opinion is not that there are three distinct species—the Barren Ground, woodland, and mountain—but that all caribou are of a single species, *Rangifer tarandus,* and that the three main varieties are all subspecies, and that there are crosses on the fringes of various ranges among the subspecies.

The Boone & Crockett Club, keeper of official big-game records, has undoubtedly worked out the most sensible system in trying to set up classes for animals from varied locations. Obviously they all could not be lumped together because of widely differing sizes of animals, and antlers, from different ranges. One category is for the Barren Ground caribou, with scientific subspecies names listed as follows: *R. t. granti* and *R. t. groenlandicus*. These heads are all from Alaska, the Yukon, and the Northwest Territories. Next comes the mountain caribou, listed as *R. t. caribou* "from British Columbia." Third is the Quebec-Labrador caribou, with scientific subspecies names the same as for the Barren Ground. Fourth is the Woodland caribou "from eastern Nova Scotia, New Brunswick, Newfoundland." The

*In late September, three Barren Ground bulls in velvet are moving toward their regular rutting area.*

scientific name for this one is the same as for the mountain caribou of British Columbia. In general Barren Ground caribou have the finest antlers and the mountain strain are the largest animals.

Probably the classification puzzle is far from finished. Hunters still speak of the outsize Osborn caribou of northern British Columbia as a separate subspecies *osborni* (sometimes *R. arcticus osborni*), and consider it a strain of the mountain caribou, which some books list as *R. montanus*. The record-book listing of the Quebec-Labrador caribou is believed by some experts to be a cross between races of both the Barren Ground and the woodland. Thus, you see, naming of these eccentric, whimsical deer is as much of a puzzle as their oddball personality.

Nor is that quite all. In the late 1800s and early 1900s some brilliant government planners decided to import European caribou— reindeer—for allegedly starving Eskimos and Indians in Alaska. A few hundred were brought in, stocked chiefly in southern Alaska. There were even Lapland herders brought along to show natives how to manage the animals. Several decades later these reindeer had exploded to a total population of several hundred thousand.

During this period residents were encouraged, ridiculously, to kill off wild native caribou, because they attracted groups of valuable domestic reindeer and drew them away. It is said that thousands of reindeer were thus lost, and that the native caribou blood was to

*This young woodland bull in Manitoba with a polished rack brushes away snow to nip the lichens underneath.*

some extent diluted by that of the smaller and less vigorous reindeer. There are many tales about mismanagement by both government and native tribes to account for the eventual debacle of the imports by the 1940s. Whatever the authentic story, happily the reindeer drastically declined and the experiment ended. Conceivably, however, there may be remnant European reindeer blood in some North American caribou still today.

## Habitat

Across its vast range there is much diversity in caribou habitat. The names given the three types describe these quite well. The Barren Ground caribou lives on the tundra where all vegetation is low to the ground. Many ranges have no trees at all. Wherever there are trees they are dwarfs only a few inches high. However, these animals move down to the tree line at times, and some of them live where there are scattered trees but seldom true forest.

Some of the northernmost herds, living on Arctic islands not far from the North Pole, spend several months each year in continuous night, and of course several more in continuous light. Here vegetation is indeed sparse, and the land a desolate, unmarked, and seemingly endless expanse.

The mountain caribou has adapted over much of its range to terrain identical to that utilized by moose in the same region. It is in fact hunted along with sheep, bears, and moose. It lives on the steep, high slopes, often ranging as high as mountain goats. As high as soil goes, the caribou is found. In this mountain terrain, of course, there are more trees than on the barrens farther north. Clumps of conifers break up the lower landscape, and there are large thickets of willow and other low growth.

The woodland caribou has adapted to forested regions, just as its name implies. Some of this variety live in mountainous areas, too,

*Slopes and valleys of the Yukon host the constantly moving herds of Barren Ground caribou. A scene full of caribou one day may be totally without them the next.*

but in general where there are trees. This is the caribou that originally was found in the mixed forests of the northern states, particularly in New England, where, especially in Maine, it was fairly common late in the last century.

## Feeding

It might seem that in their northern habitat food would become a major problem for an animal as large as a caribou. However, there are greater riches of forage than one might envision. The staple of diet for many caribou herds, and the item commonly associated with them and Old World reindeer, is the lichen usually called caribou moss or reindeer moss. It is a pale, spongy, mosslike growth of the Barren Grounds and also of much of the southward caribou range that grows anywhere from an inch or so to a couple of feet or more thick. Vast areas are literally paved with it.

This lichen tides the herds over hard winters, yet the animals relish it so much that they seldom pass it up even during the brief growing season, when other fare is available. Caribou moss, so some researchers have long believed, is one of the chief reasons for the nomadic wanderings and migrations of caribou bands. When a group is feeding it does so with concentration, each individual moving in a circle and cleaning up most of the lichens in reach, then moving ahead of its feeding neighbors to repeat the process. Thus any area where lichens are plentiful may be quickly overgrazed.

At these latitudes renewed growth is exceedingly slow. A heavily grazed expanse may take several decades to renew. Conceivably this is why caribou herds are constantly roaming, ever seeking new and lush forage, and why after several seasonal passes along a certain migration route they give it up. Their huge feet trample possibly more than they eat, and this is actually more destructive than the grazing.

Of course lichens are by no means the only fare in the far north. Summer growth over most of the caribou range is swift and abundant, and surprisingly varied. There are grasses and sedges, both staples of diet. The flats and the stream courses abound in low willows, which are eagerly stripped. Dwarf birch receives the same treatment. From midsummer on there are abundant mushrooms and varied fungi. Caribou are fond of these, although the nourishment potential is low. Blueberry bushes cover the slopes, along with crowberry, mountain cranberry, and bearberry. Labrador tea is also avidly eaten.

Basically caribou are grazers. However, the woodland variety has more opportunity than the others to accept browse, such as birch and willow. In some areas aspen also is important to them. Nonetheless, for all the races of caribou the grasses and mosses are the meat and potatoes of their menu. In winter, of course, they must take whatever they can get. The lichens now are more important than ever, and so are dead grasses covered with snow. The animals dig deep holes through the snow to get at them, pawing with flying forefeet down 20 inches or more to uncover them.

It is in winter that the long treks are made seeking food. This takes some herds from tundra to the tree line, where browse of varied kinds is available, and now of course they are forced to utilize it. In some ways caribou are wasteful foragers. Commonly they have been observed moving along at a fast walk, or even

*Caribou feed heavily on low-growing plants such as the dwarf willow (red) and the two kinds of lichens shown.*

a trot, mashing moss with their ungainly feet and lowering their heads to seize a mouthful every few steps, chewing and swallowing as they travel. When they dig in the snow and trample it, the compressed areas become hard and icy, making further effort there difficult, so the group simply goes its nomadic way.

Except when forced to feed in darkness because of sun position in the Arctic, caribou feed during daylight hours. When the period of the "long sun" comes they may be out at any time. But when full, like other ruminant deer, they bed down and chew their cud. Invariably the bedding spot during the summer months is on a high bench or an open spot with a breeze, and on a patch of snow if any are present. In such places the animals stay cool, and swarming insects, a terrible irritant throughout the Arctic summers, are inhibited.

For many years the lore has been passed along that caribou use their flattened, vertically set brow tines actually as shovels to dig away snow that covers forage. Since the palmated tines never extend, even in the largest bulls, past the muzzle, and are in a plane parallel to the face, that would be a neat trick indeed. To accomplish it with any efficiency the bull would need to stand on his head. The "shovels" of cows and young bulls are much smaller, and calves have none. Double shovels are rare. The forefeet, not the antlers, do the shoveling.

## Movements

Although well-antlered bulls in winter coat are certainly impressive, the caribou in its specialization to its environment is not as graceful on the whole as are elk and the smaller deer. It appears blocky and clumsy, chiefly because of its outsize feet. But there is a reason for these. The animals constantly travel on spongy moss, over boggy tundra and muskeg, or on ice and snow. They need both stability and utmost support in soft or slippery going. Thus nature has supplied them with feet far larger in ratio to their weight than those of any other deer, even the much larger moose.

The hoof is very nearly round. When weight is put upon it so that the two portions, or toes, splay apart, the track, quite unlike any other deer track, is sometimes wider than long. The front feet are larger than the hind ones, individual tracks as much as 5 inches long. The hind foot, when the animal walks, slightly overlaps the track made by the corresponding front foot. The dewclaws are very large and prominent and low to the ground, and in a soft medium give added stability; their use is more pronounced than among the other deer species.

There is a seasonal specialization in the feet. During the warm months the center of the hoof grows a prominent spongy pad in its center. This apparently buoys the weight in deep lichens or bogs. When winter comes, however, the sponge shrivels and hair grows over it for protection. The edges of the hooves become very sharp and hard now along the outer rim, and the shrinking of the center pad makes them more concave. The grip on ice or frozen tundra is thus improved.

When a caribou walks it appears somewhat awkward and shambling. The head may droop and the animal appears like some wildlife bum. It often stands, at rest, with head down. But when it trots it is literally regal, especially a big bull in full winter regalia. Now the head is up, the nose outthrust so that the antlers lie rigidly and unswaying along the back. Possibly because of its big feet, and the spongy tundra growth, the trotting animal raises its legs higher than other deer at this gait, like trotting horses in sulky races. The trot is swift, with long strides, and it is tireless. But when the caribou breaks into a gallop, it again appears rather awkward. Yet it is fast, going full-out probably close to 35 miles an hour. However,

*At first glance a bull caribou may seem clumsy, compared to other deer, especially with its horse-like face and blunt muzzle. But the chest and neck mane of males in autumn is striking, and the animals can move rapidly for long distances over uneven ground. This bull has not yet shed its summer velvet.*

at this gait it soon tires, often to the point of opening its mouth to gasp in more air. Shortly it must drop back to the trotting gait once more.

As already mentioned, the ankles of the caribou are built so that as they walk or run the bones and tendons make an audible snapping or clicking sound. Hunters listening intently have located a herd of animals on the far side of a low ridge by picking up the sound. During migrations of large numbers the sound can be detected at a substantial distance.

The migrations are perhaps the most notable characteristic of the caribou. Although scientists have from time to time voiced uncertainty about what causes the mass movements, it is rather well substantiated that the need for forage sparks them. In summer most caribou are seen in small bands. The woodland variety especially is more solitary than the others. It is in fall that the animals begin banding together. As food supplies for large numbers of them dwindle they begin their nomadic treks seeking fresh supplies.

While the actual reason for the mass movements may be simple enough, native imaginations have been gripped since ancient times by their sudden appearance. A tundra can be covered with moving caribou one day and perhaps empty of them the next, with the observer having no idea whence they came and where they went. Eskimo and Indian legends note well these ancestral movements. Thousands of people of the far north have lived well over many winters on the "beef of the Arctic," which is among the best of all wild game. Thousands have also starved when, lying in wait along the usual migration route when it was time for the caribou to appear, none have.

It is said that during the largest of these mass movements bands have been seen with a dozen or more on the average abreast and thousands of followers trailing out for several miles behind. Hunters have ridden mounts over 100 square miles of territory looking for a trophy bull and seen not one, only to awake one morning in camp and discover the tundra covered with caribou. The opposite commonly happens, too. A hunter after a mixed bag pays no heed to the swarming caribou, because a good one will be easy to tag after sheep, moose, and grizzly have been taken. By then, however, every caribou is gone!

Not all herds find long migrations necessary. Certain herds trade about over their local domain. A few move north rather than south. Most move southward toward the tree line if they are on open tundra. The Barren Ground caribou is ordinarily the longest migrator. A 100-mile move is nothing. Some are said to migrate seasonally 600 to 700 miles, or even farther. Woodland and mountain caribou in general make much shorter shifts. Migrations begin as early as August, there is a pause during the rut a few weeks later, and then the movement continues.

Caribou are excellent swimmers, especially when in winter coat. The underfur is dense wool, with an outer coat of long, hollow, air-filled guard hairs blanketing it. This coat is not only freezeproof but also exceedingly buoyant. When caribou come to a river, they plunge in, swimming dexterously and swiftly. They cross lakes the same way.

## Breeding

Caribou bulls put on an enormous amount of fat during late summer, to tide them over the rigorous rut. It lies in a thick sheath over the back all the way from the withers down across the rump, where it is several inches thick. Over the period of late August and on through September the animals grow their winter coats and strip the velvet from their hardened antlers. Although bands may have been composed of all sizes and both sexes, the bulls have been staying to themselves and together. But as the time for the rut arrives about late September, bulls that had been buddies a few weeks earlier are now edgy and unfriendly with each other.

Their necks are now swollen. Individual bulls begin acting a little bit silly, leaping about, racing off for no apparent reason, then pausing as if they've forgotten what they had in mind. By this time of year herds that migrate are well on their way. But the rut, which lasts overall about six weeks and in most latitudes has its peak from late September to the middle of October, interrupts everything. Whatever the intent may have been previously, bulls now have only a single concern, to gather a harem and keep it for themselves.

They cease feeding. Sparring among the bulls begins. It is the four- and five-year-old bulls that generally have the finest, largest antlers. Fighting grows swiftly until severe battles are entered. Each bull is now intermittently a wryly comic sight. Between bracing other bulls, fighting, running hither and yon disciplining his cows or trying to inveigle more, he is busy to the point of total frustration. Harems are generally not as large as those gathered by elk. A dozen cows form a large one. But the master is extremely attentive and ready to battle to retain his dominance. Some battles end in serious injury—loss of eyes, broken antlers—and occasionally in locked antlers. Many young bulls are discouraged by the older ones, however, simply by rushing charges.

Caribou are not vocal, like elk, during the rut. In fact, caribou of both sexes utter few sounds, except for occasional coughing grunts used apparently to keep in touch with each other. By the end of the rut the herd begins moving again, but the once magnificent bulls are now gaunt and bedraggled.

## Birth and Development

The bulls must now feed avidly, to lay on fat for winter once again. Within a couple of months after the rut, they drop their antlers. This is earlier than most other deer. The cows keep their spindly antlers longer, on into the spring. As winter wears on after the rut the bulls gradually get together again. Their groups, especially of older individuals, wander off from the cows and yearlings. Commonly as a migration continues the cows go farther than the bulls.

Most of the cows become irritable with their tag-along yearlings at least by April. Many simply drive them away. The youngsters then band together in herd-fringe groups, apparently able to console each other by their own numbers. By April, too, most herds are on the move again, slowly heading back to the summering grounds. As noted earlier, this is chiefly true of the Barren Ground variety. The others seldom have as far to go, the mountain caribou moving only higher up and the woodland caribou possibly out of the forested areas to more open country.

Late in May and on into June is the calving season. There is a most interesting adaptation among the caribou species regarding number of offspring. Twins are common among woodland caribou. This variety is always likely to have a less fragile and more assured food supply. Seldom, however, do Barren Ground caribou give birth to twins. Scientists believe this may be because too many fawns would build up herds faster than the tundra forage could support. Many twins, with fair survival, could double the herd size every several years. Few Barren Ground cows bear more than a half-dozen fawns in a lifetime.

Caribou are the only North American deer that seem to have little regard for where the fawns are born. Cows make little if any attempt to select a hiding place or a protected site. Wherever they are when the time arrives, they simply stop and give birth. This habit also is a specialized adaptation of nature. Unlike other fawns, baby caribou, though somewhat wobbly at first, are able to stand and run, and thus follow their nomadic parents, as soon as they are dry.

The fawns are not spotted. They are a uniform brown, with dark lower legs and nose. They are not very large, averaging 9 or 10 to possibly 12 pounds. But they are tough little creatures, able to cover many miles alongside their mothers. Some, of course, are caught and eaten by various predators. However, it is interesting that the wolf, with which caribou have lived since ancient times, seems to have little effect overall on ups and downs of the caribou population. In certain areas where wolves have become overly plentiful they are believed to have wiped out herds. But it is especially interesting that wolves are said to be unable except in rare instances to run down and kill healthy adult caribou.

Certainly calves do fall prey to the lynx, and to bears, wolverines, and wolves. But usu-

*Gray wolves are the most important natural predators of caribou almost wherever caribou roam. But wolves prey mostly on individual caribou that are easiest to catch—the calves, the old and infirm, or the injured. The strongest, most durable caribou survive.*

*In late summer the handsome bulls begin to shed the velvet from their majestic, swept-back antlers. The main beams may measure as long as 5 feet. The finest bulls have brow tines, usually called shovels, extending over their face.*

ally there are enough sick or injured or simply aged animals tagging along behind the main herd to keep predators filled. In general the calves as well as all adults are surprisingly healthy. As summer moves along, hordes of mosquitoes and black flies bedevil them. Nose flies enter the nostrils and produce larvae that go through a year-long cycle, living in the nasal passages, but these are not usually fatal. The Barren Ground caribou lives out the insect period where stiff breezes blow and snow patches may offer relief. The caribou of the forests to the south are not so fortunate and suffer more.

During summer the calves grow swiftly, quadrupling their weight at least by late August. Over spring and early summer the little calves look sleek, but from an aesthetic viewpoint adult caribou are at their worst period of the year. Great patches of the winter coat loosen and shed, or hang down, exposing the bare black skin. Soon, however, the summer coat comes in. It is much thinner than the winter one, in a shade of brown or gray. It does not stay long. Very soon the long, hollow winter guard hairs begin to appear.

With the adult sexes more or less separated during summer, all are beginning to grow their new antlers. Long-yearling calves do not form branching antlers. They produce only spikes, with no brow tines. Now once again, by August, the short summer is already gone. Individual groups that have wandered for some weeks begin the turnaround for another nomadic trek, or the mountain animals begin drifting to lower elevations. Once more preparation for the rut is underway, the prime bulls bulging with fat and beginning to stand off and gaze suspiciously at other bulls who have been their summer companions.

## Senses

It is probably not quite fair to say that caribou are not very intelligent. Nonetheless, they rather often do give that impression. Perhaps the fairer view is that they live out their lives in a sparse environment where disturbances are few and where dangers of the area are well known and have been tolerated anciently, and that therefore the animals are actually a naive product of a simple environment.

The sense of hearing is adequate, but not especially well developed, as compared to that of a whitetail deer. The ears of the caribou, which are comparatively small, seem to indicate that the animal hears only what it needs to hear, and that is not much. Unusual sounds are not often present in its domain. Hunters have stalked sleeping caribou on occasion so close they practically had to prod them awake. Bush pilots report having flown low over dozing groups that either didn't hear or weren't interested enough to get on their feet.

Characteristics such as curiosity, naivety, and perplexity are difficult to separate and evaluate when dealing with animals. Because caribou—most of them—live in an open environment, unquestionably it is their sight they depend on most to apprise them of danger or unusual presences. Yet their distance vision is only fair. And, like all deer, they respond only if there is movement. A caribou is quick to spot an unusual object—perhaps a hunter, standing in sight glassing it from some hundreds of yards. But it has no way to know what the object is, and oddly caribou in general seldom seem to react with much good sense.

One may bolt wildly at the slightest distant movement. But it may then come back over a ridge to have a look at what it fled from. Groups have been observed in the presence of danger when the animals scattered every which way, ran in circles, bunched up, stood, snorted, pranced, and some even started toward the danger to get a better look. This is obviously a sign of indecision. It may also be an indication that the simple routines of caribou existence do not require any high degree of intelligence, and that the animals don't have much.

The sense of smell is acute. Because the animal is likely to see an unfamiliar object before it can smell it, the eyes are generally used first, and the very average sight, for an open-country creature, alerts the animal to the necessity to bring its nose into focus. Yet even here the animals can be perplexing. They may spot a hunter, for example, then circle, even at a run, all the while in plain sight and perhaps in rifle range, to get around to where they can catch the scent on a crossing breeze, and even then they seem confused.

Many guides with long experience in caribou country consider them plain dumb. Perhaps so. Still, on occasion they can be most elusive, and wild as hawks. Newfoundland hunters have often had exasperating experiences with them. Hunt pressure in the sector where the hunting is done may account for some wildness. It may be also that solitary animals are likely to act more wild because they are uneasy without a crowd of their own kind around. Certainly their senses must be acute enough, since they have survived in numbers over thousands of years in a harsh environment. Perhaps a student of wildlife should consider their seemingly low intelligence and lack of sophistication as an appealing quality perfectly matching the simplicity of their lives.

## Sign

To the hunter, wildlife observer, or photographer, sign is not as important with caribou as with other big-game animals. This is because sign does not necessarily mean the animals themselves are in the vicinity. Their roaming habits preclude this. Except for migrations of some elk and mule deer bands to winter ranges, most big-game animals live out their entire lives on a reasonably compact acreage. Caribou can be anywhere.

For example, you might find several rubs of whitetail or mule deer, where they have polished antlers, and assume that the buck that made the sign was within a square mile or less. Caribou rubs may be profuse in the timber where the woodland variety lives, or in low brush on tundra locations. But the animals may be miles away. Caribou trails, in any area they inhabit abundantly, can be seen in profusion. Some ancestral trails are cut like ruts from stage-coach days left on the western prairie. Yet again, even though a trail may have fresh sign, all it means is that bands of animals passed this way. They are not necessarily in the vicinity.

In the section on movements, the feet and hooves, and the tracks, of caribou are described. Certainly in territory where other big game ranges—moose, sheep—there is no chance for even a tyro observer to mistake caribou tracks. They are so nearly round that they are unmistakable. Droppings are another matter. They might be confused even with the smaller deer, or with sheep or moose. The hard pellets formed from the more solid foods average about ½ inch long but appear in varied shapes. At their largest they are seldom an inch in length. When soft green forage is being taken, it is difficult to decide, on a range where several game species live, which is which.

Again, it is not very important. Caribou are seldom traced by signs left during their passing. The animals themselves are either there, or they are not. Abundant sign may help evaluate a good range and a general route of movement. Even the presence of abundant food, especially lichens, may help. But an old rule of the north is that whoever looks for caribou looks for caribou period. An interesting minor sign that any nature student should be aware of is the possible sighting of caribou breath. In extremely cold weather, breath vapor arising on still air from a moving band is occasionally visible before the animals are.

## Hunting

Caribou season for sport hunting traditionally opens early. The Barren Ground bulls, in August and early September, are likely to be at the tundra edge or into the trees, stripping velvet from their antlers. Mountain caribou are most often collected high above the timber, and the woodland variety in areas of open forest.

There is nothing very specialized about hunting technique. Early and late in the day the animals will be out feeding, if the rut is not in progress. It is best, by far, for meat, to hunt before the rut. Only a few days after breeding begins the meat becomes strong and fat is swiftly lost.

The most difficult part of hunting is locating animals. This may require a lot of riding. Or the country may be full of them. During the day the bulls lie up along slopes and gravel ridges and on snow patches. Walking does not cover enough territory, and anyway, hunters in these lands—except in most cases in Newfoundland, for example—are invariably mounted. "Walking up" a trophy is a tough proposition. At least one should be prepared for a long hike.

Distance, all told, is the logistics problem for the hunter. Just getting into caribou country means a long trip. Most outfitters fly hunters into base camps, from which they ride to spike camps out of which they hunt. They can change from one spike camp to another if necessary. Most hunters are after a mixed bag, although a few, in remote locations, go strictly for a trophy caribou. Walking is awesomely difficult in muskeg country, but horses used by outfitters are used to it, and on horseback one can cover 10 to 20 miles a day if necessary.

Sometimes hunters simply ride the high slopes, pausing intermittently to glass. Or they ride or climb high, then carefully glass, with binoculars and spotting scope, vast expanses of surrounding country. Much of the caribou domain is open, and the animals are easily spotted. Stalking entails one major difficulty. Most big-game animals when moving and feeding stay on a course that can be plotted or guessed at with reasonable accuracy by a hunter or guide who watches for a few minutes. Caribou may, or may not, stay on course. A band may whimsically shift direction at any moment.

The trick is to make a stalk to avoid being scented or seen, if possible. However, a spooked bull may run right back to see what frightened it, or even stalk the hunter. They are quite unpredictable. Some hunters who spot a desirable trophy in the morning keep watching until it beds down. Then the stalk is made. When caribou are moving, trying to catch them on foot is virtually useless. They go too fast. Nor is following on horseback much better. Guides invariably appraise the terrain, then ride under concealment of knolls or ridges to circle and cut the course, or even to get within range of bedded animals.

At the last place of possible concealment the remainder of the stalk is made on foot. Cover may be only low growth. It may be necessary to crawl. Caribou feeding with heads down or pointed away can be stalked a few quick moves at a time. When their heads rise or they turn, the hunter freezes. Now and then ruses work, such as hunching over and walking steadily, slowly right at the quarry. They may run around and stare but be overcome with curiosity. If one must hunt during the rut, the bulls are likely to act so totally addled and stupid that getting one is not even much sport. Although the early season prior to or at the very beginning of the rut may see some bulls with tattered velvet clinging to their antlers, the rack is nonetheless hard and can be cleaned, and the meat will be superb.

Hunters should be wise enough not to go undergunned. It's not that a caribou is especially difficult to put down, but in much of the range, at least across Alaska and western Canada, grizzlies are present. The average hunter will be out for a mixed bag, perhaps including moose and sheep. Thus heavy rifles adequate for larger game are the rule. This means, in today's world of highly specialized arms, one of the big and popular magnums. Below that, the .270 or .30/06 and comparable calibers will be next in line.

Just what the future of the caribou may be is difficult to judge. Certainly the species is not presently endangered. Some time ago it declined drastically in Newfoundland, where hunting had long been some of the best. But over recent years the animals have made a strong comeback. However, there is suspicion among some conservationists that the determined probe across the Arctic for oil and the settlements devoted to various mining and other activities indicate problems for the caribou.

As more people push into their domain, with ever easier methods of travel such as the snowmobile, more and more animals will be required simply as food. In several locations even today there is no limit on how many native peoples may take, or when. If migratory herds are blocked from their ancestral movements, or the incessant nibbling away of habitat by industry and the settlement it brings continues on an ever-growing scale, the beautiful but dumb caribou could be a casualty.

# Pronghorn Antelope

*Antilocapra americana*

The American antelope is part and parcel of western tradition and heritage. It is one of the most handsome and elegant of North American big-game animals, a highly specialized creature fitting a niche of habitat that other large animals, except the once-abundant buffalo, were not able to utilize in numbers. It is a true native, having been on this continent for millions of years. Indeed, it evolved here in several varieties long ago, has no close relatives anywhere in the world, and is the sole remaining member of its family.

Thus it is especially curious that the animal is the victim of a misnaming which has caused confusion and argument among hunters and observers of wildlife since the country was settled. Presumably early explorers and pioneers in the west called it "antelope" because it is in some physical attributes reminiscent of the antelopes of Africa, Asia, and elsewhere. But there are no antelopes outside the Old World.

The popular name stuck, until scientists discovered that the animal was not one of that tribe. To set the record straight it was given the official name "pronghorn" because of its unique horn configuration.

*Antelope have* horns *rather than antlers. They are black, are shed every fall or winter, and occasionally grow as long as 18 inches. The cylindrical tips curve backward and usually inward. A single, short, flattened prong points inward, giving the animal its name.*

121

Although attempts were made to teach the general public to use the name "pronghorn," they were to no avail. Many scientists nowadays bow slightly to common usage and try to avoid misunderstanding by using the term "pronghorn antelope." To the hunter, the westerner, and the tourist interested in looking at wildlife, however, these dashing speedsters of the plains will always be simply antelope—American antelope.

Although the antelope ranged in prehistoric times over some suitable country east of the Mississippi River, as proved by fossils discovered in Illinois and Wisconsin, pioneer Americans found none there. The range west of the Mississippi was broad, from the Baja Peninsula and much of northern and interior Mexico north throughout all habitable areas from California and southern and central Texas to Washington, the Dakotas, and the Prairie Provinces.

It is said that there were millions of antelope in immense bands almost everywhere throughout this vast range, with the concentration in its general center. Some estimates place the number at 50 to 100 million. These obviously are just guesses. Whether accurate or not, certainly the animals were supremely abundant. The incursion of pioneer settlement soon and drastically cut the population. Tens of thousands were killed for meat, to feed settlements, army camps, railroad crews, and so on. The coming of fences decimated the herds. Although perfectly capable of jumping a fence, they had never encountered such barriers and had never learned to jump. Running at high speed, entire bands were killed when they struck fences.

Vast numbers were rounded up and slaughtered when domestic cattle and sheep came to the western ranges. They were considered competing nuisances that consumed valuable forage. Numerous attempts were made to utilize antelope hides. These proved failures. The hair is brittle and loosely attached.

Hides could not be tanned with hair on. The skin itself makes a thin, stretchy, porous, and inferior leather. Nonetheless, with all such influences chipping away at what had seemed a sea of endless antelope waves across the plains, the species was brought precariously close to extinction. By early in this century, careful surveys showed them gone from most of their ancestral range. Only a meager 12,000 to 20,000 antelope remained.

A last-minute all-out cooperative effort among landowners with new understanding, lawmakers, and the new crop of game managers coming onto the scene saved the antelope. Transplants reestablished them in suitable portions of their original domain. Today under careful management and cropping to keep their numbers tailored to available habitat, a rather stable herd estimated at about 500,000 is present. Today's range, of course, is considerably smaller than the ancestral range, but remaining animals are in no danger.

Because of their specialized habitat, and man's intensive use of most of it, there is no possibility of enlarging the present antelope range. However, on the plus side, antelope have learned to live with fences, still very seldom jumping, but crawling under or through. Landowners generally tolerate them. And they are unique in having few predaceous enemies actually capable of disturbing population balance. Bobcats, coyotes, and eagles do prey incessantly and successfully on fawns. In some areas, half of the fawn crop is gone by fall. Coyotes and bobcats also kill a substantial number of adults. Pressure from those predators depends much upon the supply of staple rodents and rabbits. Except in deep snow, a coyote or group of them cannot easily run down and kill an adult antelope. The pronghorn is too fast. Antelope also fight viciously, striking with sharp hoofs to defend fawns or themselves. They have often been observed, several does together, running a marauding coyote clear out of the country.

## THE PRONGHORN

**Color:** Rich tan over upper body and outside of legs, shading reddish-tan in north to pale on southwestern desert ranges; white beneath, inside legs, and large patch on rump; throat white interrupted by a pair of wide dark tan bands; lower jaw and cheeks below eyes white; face of buck covered by black or very dark brown mask up to horns; doe with only suggestion of darker color on face; nose, both sexes, black; buck with black neck patch on upper throat and rear of lower jaw, also with a few long, black-tipped hairs on mane; short tail tan above.

**Measurements, mature bucks:** 3 feet or a bit more at shoulder; overall length 4 to 5 feet.

**Weight, mature bucks:** 100 to 120 pounds, exceptional specimens a few pounds heavier.

**Horns:** Black; 10 to 15 inches outside measurement, rarely to 18 or more; tips generally rather cylindrical, curving backward and a bit inward, but this is highly variable, some tips curving forward or directly inward; a single usually concave short, flattened prong on front of each horn about midway, thrusting forward and up; horns from skull to prong flattened laterally; normally horns appear to slant slightly forward.

**Does:** 10 to 20 percent smaller; with horns, but inconsequential, seldom more than 4 inches high and normally without prongs.

**General attributes:** Graceful, handsome, beautifully proportioned; unique ability to flare white rump patch broadly when disturbed or angry; astonishing speed when running; possibly world's fastest animal over substantial distance; phenomenal eyesight in treeless plains habitat.

## *Range of the Pronghorn*

*A mature pronghorn buck is generally 10 to 20 percent larger than a doe. Most bucks weigh 100 to 120 pounds*

A severe blizzard with deep snow is one of the worst enemies of the pronghorn. They may be forced to drift before it, and they may pile against fences, smother by lying down too close together in concentrations, or starve because they cannot paw down to forage. However, the animals are remarkably adapted to withstand severe cold. The pithy, hollow hair contains large cells of air. Although the fine undercoat is rather inconsequential, a highly specialized set of muscles in tissue under the skin allows the antelope to raise or lower the main coat hairs, adjusting the angle to the weather. At zero or below it is held close and flat, forming perfect insulation. In hot weather it can be raised and slanted to take advantage of cooling breezes.

Considering the speed at which antelope are able to run, and the rough, often rocky terrain they encompass, one might suspect that accidents would take a high toll. This is not the case. The animals are astonishingly adept at maneuvering at full speed over hard and uneven terrain. They have an uncanny sense for avoiding deep mud or quicksand. Although the legs appear fragile, the bones are highly specialized for adaptation to rough usage. They are unbelievably strong. Arthur S. Einarsen, in his book *The Pronghorn Antelope and Its Management,* tells of an experiment he conceived and had carried out to determine the maximum load weight each leg bone could bear without breaking, as compared to the leg bones of cattle. The results showed that the foreleg bone of a domestic cow, which is seven times greater in weight than an antelope, was crushed at 4000 pounds less pressure. The antelope bone withstood over 45,000 pounds per square inch!

There are some records of accidental deaths, of bands racing over a cliff in a snowstorm, for example. By and large, however, aside from predation upon fawns, disease and parasites which weaken the animals are undoubtedly the chief causes of mortality on the range. Antelope are not, however, particularly long-lived. An animal five or six years old, or at most eight, is on the downhill side.

The two most emphatic facets of antelope personality are extreme nervousness and curiosity. The nervousness does not seem to stem from constant fear, but from a super-acute

*Coyotes exist throughout pronghorn range, at times in goodly numbers. They prey on fawns before the fawns are able to run fast or far. They will also kill adults weakened during severe winters with deep snow. But coyotes ordinarily do not have great impact on pronghorn populations.*

alertness to everything that occurs around them. Their eyes are ever scanning. Some imaginary danger may put them briefly to flight. Or a sudden impulse will send a band or a lone pronghorn racing off, circling, standing to pose atop a ridge. An almost constant movement is the way of their lives, inherent in the very nature of the terrain. Even sleep is fitful, a few moments of dozing, then the head comes alert and the big eyes scan the surroundings.

Any unusual sight arouses overwhelming curiosity. A scrap of paper blowing across a plain may have several antelope stalking it with prancing walk or trot. This strong curiosity has got many a pronghorn into trouble. Hunters since the early pioneers have used ruses such as tying a cloth to a bush so that it flutters in the wind. Hiding nearby, they wait for a curious antelope to investigate. Lying down and waving a hat gets the same kind of attention. Some individuals may come a mile to find out what is going on.

The type species, *Antilocapra americana americana,* makes up the bulk of the antelope population. Four subspecies are recognized, but they are not very important to the casual observer or hunter, and not numerous. One is the oregon antelope, which presently resides in extremely modest numbers in the sagebrush country of eastern Oregon. The Mexican pronghorn is a southwestern subspecies found on plains and deserts in parts of west Texas, New Mexico, and Arizona. The Sonora pronghorn is an endangered desert subspecies of the Mexican state for which it is named and of a small area of southern Arizona. The Peninsula pronghorn is found in Lower California. The four subspecies make up certainly no more than 50,000 in aggregate, if that, of the estimated 500,000 animals.

The center of antelope abundance within the entire range is in Wyoming. This state contains perhaps a fifth of the total continental herd. Montana is second to Wyoming. Sustained antelope abundance in this general two-state region indicates that here is located the optimum type of range. Other states and provinces with antelope offer proper natural habitat, or proper habitat not denied to the animals by human changes in land use, only in modest amounts.

## Habitat

The pronghorn is a product of the wide-open, treeless country of the west, a creature of the rolling plains and grasslands, and of the sagebrush flats and lower foothills. The preponderance of the very best range, and most of that in regions outlying from the optimum center, is not, as it has sometimes been described in romantic western literature, any endless sea of waving yellow grass. Indeed, much antelope habitat is quite rugged. Barren buttes and plateaus thrust up from the undulations of the lower country. Rocky outcrops show through the low vegetation.

In northeastern New Mexico, for example, where some of that state's prime antelope range is located, the land is "malpai" country, where grass and weeds and low brush grow from among jumbles of "bad rock," the black hummocks and ridges spewed eons ago from extinct volcanoes. In the Big Bend Country of western Texas, antelope habitat is in the slanted, rocky desert valleys and low ridges between desert mountains.

Sage and antelope are a duo over much of the animal's territory. There may be flat dry lake beds on which bands wheel and play, or flat expanses, but the arid ridges with hard, stony soil and sage generally will not be far away. Just as the mule deer seems unable to colonize flat lands, the antelope cannot abide the confinement of forest. It must be able to see for long distances in order to feel secure.

The distribution of antelope has always been limited by two types of barriers. One is

*When Americans first began migrating westward, they found antelope by the millions shar-*
*ing the land with bison. The bison are almost gone, and antelope now live with them on a*
*few scattered refuges. These antelope were photographed in central Wyoming.*

stands of heavy timber. Antelope refuse to enter them. As an aside, it is interesting to observe that when antelope are coming to a waterhole, for example, and tall sage thickets of 6 feet or so are present nearby, the animals invariably skirt them. Enemies might lie in wait there, particularly at a waterhole, the gathering place in an arid country for all wildlife. Besides, while passing through the tall sage, the antelope cannot see far enough.

The second barrier of terrain is what are generally called badlands. These are severely eroded areas. In numerous instances where excellent range was present on one side of such an expanse, antelope have failed to find and utilize it, yet they have drifted along the edges of the region distantly to populate a range of inferior quality.

Even though heavy timber acts as a barrier to distribution, in certain parts of their range antelope have been forced into quasi-timber areas by the encroachment of man and his land uses upon their preferred domain. These habitats are not forest or woodland in the literal sense. But they are the fringes of it. In Arizona, for instance, antelope presently utilize country of scattered juniper quite unlike the prime habitat of eastern and central Wyoming. The same is true in Texas and New Mexico. In part of the Oregon range they were also forced by human pressure into stands of juniper. It may well be that originally antelope utilized authentic pure grasslands far more than they do today, even on the better ranges. Piñon, juniper, and varied brush and weed associations now are utilized here and there, not always by choice but because that is all that remains or because of man's influences upon the original grasslands.

The span of elevation within antelope habitat is broad. But it appears to have its most definite limitation on the lower side. Only in the prairie provinces of Canada do pronghorns find congenial altitudes of only 2000 to 2500 feet. Elsewhere over virtually every state where they are found the favored altitude is from 4000 to 6000 feet, except in locations such as the Dakotas and Baja California, where they occupy an average altitude between 3000 and 4000 feet. Whether or not altitude is a fact or influence related to abundance is not known, but the level at which pronghorns are most abundant in range center is from 5000 to 6000 feet. Pronghorns move higher than that, however, to 8000 feet or more on certain open mountains.

All of their habitat is arid in varying degrees. But it is never waterless. There have been suggestions in some writing that antelope can live without water, utilizing that from their forage. This is untrue. The animals can get along for several days without water if necessary. But the waterhole is a hub of any pronghorn habitat, even though it may not be visited without several miles of travel, or every day. Wholly waterless country cannot sustain pronghorns, although some astonishingly barren-appearing places with seemingly scarce forage can.

## Feeding

The antelope is a very nervous animal. Its feeding habits reflect this. An animal may walk along taking a nip here and another there, suddenly race off at a run for half a mile, stop abruptly, pick out a certain weed that seems appealing, eat several bits, then perhaps wheel and race back to where it was to start. Possibly because antelope quite literally have perfected the art and science of running, thus expending much energy whether disturbed or simply racing around in play, they feed over longer periods than do deer. And they may feed at any time, day or night.

Researchers doing stomach analyses on antelope have noted the unusual fact that with few exceptions the stomachs are totally full. This would seem to indicate numerous feeding periods around the clock, with only short rests in between. At one time landowners believed that antelope were a scourge on the range, competing with cattle or sheep for grass. Scientific studies long ago proved differently. Browse plants are the mainstay of antelope diet.

Wherever sage is available—which takes in much of antelope territory—it is one of the most important items of diet. In winter especially, when other plants are not available

*A male pronghorn is a graceful, handsome, well-proportioned mammal with the unique ability to flare its white rump patch when frightened or disturbed. It has phenomenal eyesight and is among the world's fastest quadrupeds. Mature bucks average between 100 and 120 pounds, and stand three feet tall or so at the shoulder.*

in abundance, sage of one variety or another is a staple. Although the palatability of sage may not be high, studies have shown it to be an important protein source, and to contain more carbohydrates and much more fat than such rich cattle feeds as alfalfa. Common additional browse plants are rabbit brush, saltbush, juniper, and bitter brush. There are numerous others.

Next in importance after the browse plants, which are eaten all year, are various weeds, or forbs. These grow in antelope habitat in infinite variety. Sour dock, chicory, dandelion, Russian thistle, mustard, wild peas, lupine, mullein, larkspur, various clovers, and even wild onion and locoweed are examples. Grasses are taken in the least quantity, and chiefly in spring when green shoots appear. The ratio overall is roughly two-thirds browse, as high as one-fourth weeds, the remainder grasses. This ratio, of course, differs from area to area, depending on what is available. In some instances browse may make up as much as 85 percent of the forage.

It is apparent that pronghorns are perfectly fitted to their environment even in their digestive system. Compared to bulk of food eaten, there is very little body waste. They also are what might be called conservation-minded feeders. Most large animals bite and pull at the same time, thus tearing out the roots of low-growing forage plants. Antelope are almost dainty in the way they snip each stem with a clean bite. This allows a reissue of the plant and very little plant destruction by the animals on any range. This is important in public relations, since they must in our modern day share range with cattle, which quite oppositely are accomplished root pullers.

Although antelope range widely while feeding, and may be active at any time, their loose routine is to move about early in the morning, then take a rest for cud chewing for an hour or more. But there is usually a second feeding period following this. If the day then becomes hot, the animals lie down during the worst of it. A late feeding period then begins as afternoon wanes. However, whimsical and nervous as they are, bands may be up feeding and moving at any time.

Because present-day antelope range is almost all in multiple use, either privately owned ranchland or else federally owned land used for grazing, the welfare of the animals depends almost entirely on their acceptance by cattlemen and sheepmen. Occasionally where crop fields such as alfalfa border antelope range, the crop makes up part of the diet and the animals become exasperating pests—not, however, from their feeding on it, which does little real damage, but from their constant jittering, running hither and yon through it, lying down to flatten one patch and then nervously jumping up and moving off to try another spot.

As far as open rangelands are concerned, however, it is a curious fact that even large numbers of antelope seem to have little if any adverse effect upon it. Where cattle are excluded from a range used by numerous antelope, native grasses instantly begin to thrive. On seriously overgrazed lands, antelope get along well while cattle must be removed. As soon as the cattle are taken off, the range begins to restore itself. It is this modern understanding of the feeding habits of the pronghorn gleaned by science, that has gained the cooperation of landowners in helping to keep the population at a stable level. Interestingly enough, certain areas in the southwest capable of producing only the sparsest plant life cannot sustain cattle, yet are utilized by antelope.

## Movements

The antelope, one observer has said, is motion at its best. The wide-open habitat with its endless distances seem to invite wild, free, and swift movement. The antelope is so specifically attuned to it that it evolved as possibly the world's swiftest runner. And it runs with unbelievable grace and flowing motion. A band running together is reminiscent of a flock of wheeling birds, swirling with evenly flowing motion as if it were one individual.

At 20 miles per hour antelope are barely at a good full trot. They can run for miles at this speed. At 35 miles per hour they hit what might be called a fast cruising speed. They have been clocked at this pace over a course of several miles without showing the least sign of tiring. A bedded antelope can catapult into the air when disturbed and land running. The gait is not a series of bounds such as deer

make. It begins as a kind of trot, left front, right hind, and vice versa, but each foreleg reaches far out and the animal is almost instantly revved up to 30, then progressively to 35, 40, 45, 50.

It moves at the higher speeds only when chased or encouraged, but it still has lots left for serious occasions. Whole bands have been clocked running together at 50 and above. Probably 55 miles an hour is average maximum. However, there are observer records touching 60, and some believe individual animals under proper conditions may be capable of 70 for a short spurt.

*Pronghorns run with such fluid, graceful motion that they may appear to be traveling slower than they are. These animals are cruising at about 35 miles per hour but could accelerate to 55 miles per hour if necessary.*

The pronghorn has several special adaptations to make possible such speeds. The front feet are larger than the hind feet, in mature bucks close to 3 inches long, as compared to about 2½ inches for the rear feet. It is the larger forefeet that hit the ground hardest when the animals are running, and that support most of the weight. The lungs are extra-large, and the heart is double the size of that of animals of comparable weight.

There is a nerveless cartilaginous padding on the hoofs, particularly thick on the forefeet. This serves to cushion the strike of the feet at high speeds. Because of it, antelope seldom show any lameness or tenderfootedness regardless of the rough character of the terrain. There are no dewclaws, which conceivably might be a hindrance to swift movement on rough or rocky ground. Further, the design and fitting together of ligaments, tendons, and bones of the lower leg are so perfect that it is virtually impossible for the animals to suffer sprains or other related leg injuries. This is one reason pronghorns are such miracles of swift flight over broken terrain.

One of the most unusual specializations for speed over substantial distances is found in the windpipe. When a pronghorn runs at high speed it always does so with mouth open. This has fooled some observers into believing the animal was tiring or out of breath and gasping. It is gasping all right—pulling in great drafts of air that its nose could not accommodate. The windpipe, however, easily accommodates this large intake. It is oversize, indeed twice as large as that of several animals of double the weight.

Antelope are notorious for their irresistible desire to race any moving object in their domain. This trait was noticed by early pioneers on the plains. Spotting a galloping horseman, a moving vehicle, a train, they commonly run on a long slant toward it, eventually come alongside, then appear to take great joy in speeding up not only to outdistance the competitor but also to bound across in front. At 50 miles an hour an antelope may suddenly spurt ahead and make a leap of 20 to 25 feet barely ahead of a moving vehicle.

No one is quite sure what the motive is for this. Perhaps it is really a form of race, of enjoyment. It may also be a quirky escape idea, a feeling that safety lies not only in outdistancing an "enemy" but in getting on the other side of where it was first sighted.

Although the antelope can make long, low, vaulting leaps, such as across a ranch road, as noted earlier it has never learned to jump vertically. It is easily capable of clearing any ordinary ranch fence. But in its bailiwick it has never had to leap *over* anything, and so simply never learned. A band often runs right at a fence of several barbed-wire stands and hardly slows. Each animal ducks through between strands or else underneath the lowest one. The loosely attached hair is raked out in an explosion drifting on the breeze and on the animals go. On some ranches there are locations where a woven-wire fence corners with or meets a stretch of barbed-wire fence. Antelope often wear a trail along the woven wire, straight to the spot where the barbed wire begins. There they duck through.

Yet indeed they can jump. One observer in Wyoming, trying to get a buck antelope out of an alfalfa field surrounded by woven wire, chased it with a vehicle until it became so tired it lay down. All the time there was an open gate for it to go out through—the way it came in. But curiously, when pressed and running, bands commonly pass an open gate time after time as they circle a field, somehow dubious of moving through it, or else too intent on running to notice. At any rate, when jumped from its resting place again, the buck bolted straight at the fence, sailed over it easily and beautifully, then ran up to a ridgetop and posed almost arrogantly, looking back.

How far antelope move in a day, or a season, depends on several factors: the amount of forage readily available; the proximity to water; disturbance; and how closely they may be confined by fences. It must be remembered that in modern times on ranches here and there bands of antelope are inhibited from movement because they were "fenced in." A new fence is built splitting a big pasture. Antelope in one section may have water and food, but will live out their entire lives happily enough, with only modest movement possible. They are trapped.

Generally speaking, however, the individual range of antelope bands where there are no barriers over large expanses is much broader than that of other horned and antlered animals. This is simply because movement to them is not labored. They are the most easily mobile of all hoofed creatures on the continent, and are mentally adjusted to wide movement. Nonetheless, a band or a big lone buck may live for weeks or months covering no more than a square mile or so. It will water at the same location, feed on a favored flat, bed down on a favored ridge, pass time after time through the same saddle of a narrow valley between ridges.

Even sometimes undue disturbance, as during hunting season, won't drive a band out of the country. It will become awesomely wild, running at sight of a man or vehicle a mile off, but next morning the group may be in the same area. If there is an especially rough piece of country, with snug valleys and high, rocky ridges, bands will commonly move into those when harassed out in the rolling country, coming out only to water.

The trek to water is usually made only once a day. Occasionally it may not be made every day. Much depends on the temperature. However, the daily trip is the rule. Individual bands of animals may differ in timing of the water trip, and it may differ in varying latitudes. A group of archers who hunted several consecutive seasons in Wyoming and set up sage blinds near a waterhole on a small creek observed that without fail bands there came to water in the middle of the day.

Although antelope groups like to hang around old dry lake or pond beds, sometimes because they serve as salt licks, seldom do they stay long at a watering place. In an arid expanse it is a magnet for all creatures, prey as well as predator, and thus dangerous. Coming to water, the animals often stand on a ridge and look things over carefully, then move skittishly down toward the water. They jitter about, finally move to water's edge, lower their heads, maybe flush suddenly like a flock of quail, then move gingerly back. Once they decide all is safe, they drink. With little dilly-dallying, they then prudently leave.

The uniquely specialized white rump patch of the pronghorn is related to movement from danger. As mentioned earlier, all of the pelage can be erected or laid flat by an intricate web of muscles in tissue below the skin. The rump patch is so well endowed with these muscles that every hair can be individually flared. The white rump hair is almost twice as long as the body hair. When wholly flared and looked at from the rear, the patch protrudes past the basic body contours at least 3 inches. The patch thus appears at least twice as large as when the hair is laid flat. Presumably this flaring is a signal to others in the band, or to any other antelope within sight range—which may be a couple of miles—that something is amiss. It's time to run.

This signal is known to everyone who has seen or read about antelope, but most casual observers are unaware of what occurs previous to this final and obvious danger signal. When puzzled and suspicious, first the animals raise the tan hair along the back. The bucks also erect the 3-inch black hairs of the mane. Pronghorns may begin to walk or trot slowly away with a stifflegged gait as the back hair comes up, meanwhile flaring the rump patch. Then usually they explode into a headlong run. Once the band comes to a stop, out of danger and probably several miles away, it may pose on a high place to look back, and all the animals make a quick shaking motion, smoothing all the flared and erect hairs back into place.

After this antic they are by no means as highly visible.

Scientists studying antelope have long been convinced that among their important movements is authentic play, not just among fawns but among adults as well. A lone pronghorn may make a sudden dash at a group, even some that are lying down. It races around or past. They instantly are on their feet, giving chase, circling, racing. The group may fan out, then some individuals will cut through the running pattern of the others. Such displays are common, are not related to sexual motivation, and appear to be thoroughly enjoyed by the group.

The so-called migratory movements of antelope are commonly misunderstood. There really are no true migrations in the strict sense. In winter antelope gather as a rule in large bands. They may drift some distance to a wintering area. But the movement is not necessarily routine, season after season. And it bears no relationship to latitude. That is, one band may move a few miles north, another south. Drifts from one feeding ground or waterhole to another occur at any time. So do switches of altitude. Weather and seasons may influence all such movements.

In a very few instances, in the northern fringe of range in the prairie provinces, for example, a wintering ground may be distant. This trek occurs because of deep snow which forces a long move to forage. One of the most interesting winter movements in the north is to high altitudes. The animals do not seem bothered by low temperature and bitter winter winds. When a valley fills with snow, they move right up atop the highest windswept peaks or plateaus in the area. Here forage is swept partially clean of snow by the winds. Antelope do dig through light snow occasionally to uncover food, but it is not a routine habit.

There is seldom need for antelope to swim. They are known, however, to cross large rivers now and then with no hesitancy. The coat with its air-filled hair is buoyant, and they are strong, calm swimmers. They do avoid wet bogs and sinkholes, seeming to know that they may get caught in such places.

## Breeding

Some other antelope movements are related to the fall rut, to breeding. On most ranges late summer is a dry time; water is scarce and numerous bands may be forced to gather at the same watering places. These concentrations may be congenial enough until mid-August or early September. Then mature bucks which have been hanging out with groups of does and fawns, as well as old loner bachelors and small groups of bucks hanging out together, begin to get restless. Soon scattered individual bucks pass into what some observers have called their "crazy" period.

A big buck in perfect physical condition may stand apart from a group, listless and with head hanging. He flares his rump patch now and then and looks more alert, and an incessant tremor of the rump hairs is noticeable, caused by twitching of the muscles that erect the hair. He may suddenly make a wild sideways leap, or race in a quick circle. To some extent this is reminiscent of the ludicrous antics of bull caribou as the rut begins. These weird antics of the buck draw the curious attention of the group.

Soon, however, other bucks are going through the same personality change. Curiosity wanes in the others, and now the forerunner activity of the mating season begins to speed up. Mature bucks that have been hanging around with does become jealous of them. Other adult bucks race frantically around urging does to join them. This is the period of harem forming. Bucks dash madly about vying for does. They get together a small band—three or four to a dozen—and are violently possessive.

It is wryly comical to observe the dilemma of a buck with a fair-sized harem that is being challenged. Another buck without does comes tearing in toward the flock. The herd buck gives chase. Eyes popping viciously almost out of their sockets, rump and black mane hair upended, he closes on the interloper. About then he thinks twice. What's happening behind him? Will still another buck sneak in and steal his harem? He charges back, only to be followed by the other buck. Off he goes again, determined this time to appear as formidable as possible and run the other clear out of the country. There are again second thoughts. Back he races to his does.

During this period of forming and attempting to control small harems, movement is frenzied. The dry flats are constantly inscribed by myriad floating dust plumes as buck after buck flurries around his group of females or chases other bucks. Now and then there are battles. Most of these amount only to bluffing, some shoving, an occasional severe charge during which heads come together and, usually, horn prongs catch as guards to fend off any serious damage. Sometimes horns are broken or knocked off. The horn tips, being curved back or inward or forward, cannot cause any notable wounds, but the often-sharp prongs might. Very occasionally a buck is badly hurt and bleeding, or put down and slashed with front hoofs until he is mortally wounded. It should be emphasized that this is uncommon.

*In summer pronghorns group into congenial bands, close to sources of water. But by late summer the congeniality dissolves and the "crazy" period of the rut begins. With the onset of the rut, the bucks, especially, become restless and pugnacious.*

Timing of the mating season differs somewhat from latitude to latitude. In general it occurs during September and October. The peak activity is rather brief. Antelope does, it is believed, have only a single short mating period, unlike deer. Thus if one is not bred when ready, or does not become pregnant from a mating she will be fawnless that year. This belief seems to be substantiated by studies which show that in spring all but a very few fawns are born within a period of a couple of weeks, whereas late fawns are common among deer, and the fawning period of deer stretches over a month or more.

The frenzied breeding period wears all the animals down. But it is soon completed and the fierce competition and incessant running cease. This volatile several weeks has long been a problem to game managers trying to set a proper hunting season. Most are set in late September or occur in October or even into November. A few states wisely launch their season in late August. This gives hunters a chance to collect animals in the primest condition and with meat the best. And it probably does little harm to the breeding population because of the harem-gathering habits of the animals. The problem in a late season is that the horn sheaths (discussed below) loosen soon after the rut. Many a late-season antelope hunter has dropped a trophy only to have the horns fly off when it hit the ground.

## Birth and Development

Virtually all of the does are bred the first time when they are long yearlings—that is, only two to four months past one year of age. This well may be a specialization to keep population level high, for antelope, as we have said, are not long-lived. Few live past seven or eight. Although mating is earlier in general than for deer, the fawns are born at about the same period of spring, in May and June, depending upon latitude. This means that the gestation period is long, approximately eight months.

The fawns are very pale, grayish rather than tan, and weigh only 4 or 5 pounds. They seem to be all legs, and stand only 16 inches or so at shoulder height. They develop with astonishing swiftness and are able to run within a few days, if necessary, up to 20 or more miles an hour. However, for at least the first week they lie flat most of the time, hiding in cover. The young does give birth to a single fawn as a rule, but twins are common thereafter, and triplets are not rare.

Fences nowadays prohibit the does in some places from selecting a kidding ground that they might move to if the range were open. Thus many observers have been led to believe that the young are simply dropped anywhere the doe happens to be. Given a free choice and proper variation in habitat, however, this is hardly the case, and the selection of a fawning place is uniquely interesting. Invariably they will select a small basin or valley with ridges nearby, and with vegetation about a foot high.

The fawns—or kids—are born down in the lower area. If there are twins, the doe gives birth to one and then moves away, sometimes several hundred yards, to bear the other. This instinctive plan is an aid to protection of the young, or saving one if some predator finds the other. After the births she drives off the youngsters and then leaves, commonly moving as much as a half-mile away, but invariably up on the ridge. Again no doubt instinctively, she has selected the low vegetation for giving birth so that now from her higher vantage point she can watch closely over her offspring, whereas if they were deposited in high cover she would be at a disadvantage.

The young are believed to be practically odorless, and they lie tight to the ground. The doe comes often to let them nurse, but immediately leaves to take up her watch from above. By the end of the first week the fawns

*Antelope does have one or, occasionally, two fawns, which are pale gray-tan. For the first week or so the fawns lie flat or hide motionless in cover. But they develop astonishingly fast and, if necessary, can soon run along with the mother at speeds to 20 miles per hour.*

are able to follow their mother, and by two weeks their white rump patches have developed and they flash them in flight just like the adults.

Like deer, antelope are only modestly vocal. The fawns bleat on occasion, their voices high-pitched and quavering. Does may respond with a low blatt, but this is unusual. A wounded buck has been heard to utter a deep, short blatt, again an unusual utterance. The only common sound antelope make is a kind of snort through the nose when they are either disturbed or aroused.

While the fawns are growing during the summer, yearling bucks and does are beginning to form their first horns. Pronghorn horns are not similar to either the antlers of deer or

to the true horns of cattle and the true antelopes and other horned animals. The handsome pronged horn that is seen as a trophy by hunter or observer in summer and early fall is actually only the outer sheath. Inside it is a core fed by a circulatory system. The sheath is rather lightly attached to the core. After the rut—and sometimes accidentally during it—the outer sheath is shed.

Thus the antelope horn is a kind of in-between type of growth. Antlers, for example, begin growth with an outer covering of velvet, which is really a blood-vessel system on the outside. At full growth the blood system dies, shrivels, and is rubbed off; the antler has hardened and is eventually totally shed, and a new growth begins the following spring. A true horn has a core supplied with blood vesssels that nourishes growth throughout life of the outer bony portion, which is never shed. The outer sheath of the pronghorn horn is composed of a hairlike substance fused into a solid, but not truly bony, mass.

Before the outer sheath is shed in fall, a new sheath is already beginning to sprout inside, at the tip of the core. The core is smooth and short and without any prong. Early in the year—by January or February—the new sheath sprouting at the tip of the core is 3 inches or so in length. The manner of growth is now unique. The tip is hard and shiny, like a regular horn tip. The continued growth, however, is downward toward the skull. In addition to the bony tip, the outside of the core is covered with a membranous material which covers it down into the hair at the base. The hair surrounding the core at its base also is growing swiftly as the sheath grows. As the new sheath reaches downward, it eventually covers and fuses with this hair. The complicated process is generally finished by midsummer. The new

horns at that time and on through August are unmarred and in their most handsome state.

To give an idea of how large antelope horns may grow, the longest in the record book are 20⅛ inches right, 20 left. The animal was killed in 1899. The preponderance of records—which of course do not depend just on length—run in length from 15-plus to 19-plus. The older a buck, with some exceptions, the larger the horns. But because the animals are relatively short-lived, there is a built-in limit. Further, cropping by hunting to keep herds in line with available range tends to skim off the larger bucks each season. In most areas nowadays a 12-to-25-inch horn is considered a good trophy. Nonetheless, for hunters who believe that chances at a trophy that might make the book are presently meager, a check of records will be encouraging. Over a recent six-year period slightly more than one-third of the entire 230-plus record-book heads were taken, some of them placing up toward the top.

## Senses

Sight is the all-important sense of the pronghorn. The size of the eye has often been compared to that of the horse's eye. It is larger, even though the antelope weighs only a fraction as much. The eyes are so placed out at the sides of the head that the animal has extremely wide-angle vision. It is even capable of picking up movements behind. Even though the eyes bulge, they are extremely well protected by the skull design. Researchers have even found it necessary to chip away the bone above the eye in order to remove one.

Some scientists believe the antelope eye has magnifying power. It is often likened, whether provable or not, to an 8-power telescope. It is a fact that antelope detect movement of very small objects several miles away. The high specialization of the eyes is of course an adaptation to the open habitat. This super-keen sense, coupled with the just as highly de-

*This pronghorn skull shows the bony cores, one covered by its outer sheath, the horn, which is shed every year.*

veloped running ability, forms the combination upon which the well-being of the animals almost wholly depends. Add to this the fact that pronghorns—except for occasional individuals—consort in bands and are incessantly and nervously scanning their domain. With so many eyes watching at once, the odds are high against close approach and surprise by any predator, including man.

Scenting ability is well developed, but by no means to the degree that it is, for example, in whitetail deer. At modest range, antelope pick up scents of danger on a breeze. But they have no real need for distance scenting ability. Hearing is certainly acute, but again, the source of any sound an antelope hears has invariably already been detected by the eye.

## Sign

Signs left by antelope are not very important, so far as using them to locate the animals is concerned. Certainly tracks around a waterhole or on a dry lake bed give some indication of animals present. But the point is, simply scanning an area either with unaided eye or binoculars soon locates the animals themselves, if they are using a given range. A band lying down on a hillside is difficult to spot at times. It blends well, but the white portions, which may appear to be pale rocks at a distance, catch the observer's eye and are quickly turned into antelope by use of binoculars.

As noted earlier, antelope have no dewclaws and thus their track prints simply show the two halves of the hoof. The forefoot track measures from about 2⅞ to 3¼ inches in adults, the smaller hindfoot track about 2¾. The rear of the track is a bit broader than that made by deer. However, on some antelope ranges mule deer utilize the same feeding areas and waterholes and distinguishing positively between the two tracks is difficult if not impossible. Much depends on the experience of the tracker, and of the medium in which the prints were made.

Droppings also are rather similar to those of deer. Some of them may be smaller, about ¾ inch, but many mule deer pellets are of the same dimension. When utilizing soft forage, droppings are a soft, irregular mass. This type is much like the same left by deer. During the summer when a combination of foods is being eaten, a more distinctive type of dropping is left. It is a rounded, elongate mass with pellet formation apparent.

The most distinctive sign left by antelope are scrapes made by the forefeet, in which the animals urinate and leave droppings. A small scraped-out place is dug with the sharp hoofs and body wastes are deposited in it. This is such an ingrained trait that individual animals leave numerous such scrapes daily. The problem in making use of such sign is that pronghorns move around so much. An undisturbed group may stay within 600 or 700 acres for days at a time, but it may visit practically every part of that area during an hour or so.

## Hunting

In modern days the sport of pronghorn hunting has often been badly abused by the use of vehicles to chase the animals and to put a hunter within "flock shooting" range. In almost all instances this is illegal, and many game-department people really crack down on hunters who stoop to it. Although antelope hunting would seem to be a simple process, since the animals are easily spotted, it is actually one of the most challenging of endeavors when approached in a sporting manner.

Shots at antelope tend to be long, but there is a pronounced tendency among numerous hunters to try them much too long, out to 500 yards or more, and even at running animals. It is far more sporting, and more productive for that matter, to employ craft and stealth.

One common method is to cruise ranch or public-land trails in a vehicle, watching and glassing for distant bands. If a hunter is willing to be patient, this method will often locate a real trophy buck. And a couple of days of scouting may pin down the fact that the buck is using a certain basin or general area as a home base. This method is possible only where hunters are not numerous and chousing the bands around. The less disturbance the better.

Once a desirable specimen is located, an approach must be planned. There is little cover in antelope territory. But a small draw or a low ridge or series of ridges may form an avenue for a careful hunter to close in to rifle range of 200 to 300 yards. The exasperating part of this is that a stalk that requires much time may find the animals long gone before its climax.

Sometimes it is possible to use a vehicle or even another hunter as a ruse to hold antelope attention. A driver cruises slowly along and the hunter drops out on the off side in some small depression. The vehicle goes on and while the band watches it, the hunter, taking advantage of every small bush and depression, crawls along making his stalk. On occa-

sion ranch or other vehicle trails may curve around so that a vehicle will start a band racing off in the direction of the stalker. In this case the driver—in most states it is illegal to drive off-road while hunting—tries to get on over a ridge out of sight, hoping the animals will stop running by the time they get into range of the hunter.

Although scores of hunters shoot at running antelope, sometimes firing thirty or forty rounds without collecting a trophy, running shots should be avoided. It is easy to wound animals this way, or to shoot the wrong one in a band. Standing shots should by all means be the goal.

Experienced antelope hunters check out a given expanse of hunting country, seeking the roughest portion. It is here, in the tight basins and narrow valleys between steep and often high ridges, that many an old loner trophy buck will hang out. It is here also that every antelope from flatter country in the region will head when a swarm of hunters begins combing the flats. Many mill-run hunters used to vehicle hunting don't want to tangle with the rough places, so the hunting isn't crowded.

*Often a hunter has only a split second to pick out the best buck, or the one trophy buck, if any, in a herd. The animals are always off and running when man appears on the scene.*

Further, these roughs lend themselves best to a careful stalker. Moving toward a ridge crest, then lying down and creeping up to peer over may put a trophy in the sights below within easy range.

It may or may not be legal in the state where you hunt to take a stand within range of a waterhole. If it is legal, a small blind of sage set up on a ridge perhaps 100 yards distant from a waterhole known to be used is a productive plan. Stands may also be taken beside a fence near an open gate where bands habitually pass, or near a much-used fence crossing. A small depression beneath a fence may be used daily by a band moving from one pasture to another. Their sign—scraped-off hair, and tracks—will indicate usage.

A stand between a watering place and a series of ridges and roughs, if properly planned, can also be effective. Perhaps there is a deep saddle in an otherwise steep, high, long ridge. The waterhole, let us say, is half a mile or so west of this north-south-running ridge. Animals from the roughs behind the ridge will stream through the deep saddle as the easiest path to water. This movement will probably be sometime during the middle of the day. Given a breeze from the east, a stand taken along one spur of the west side of the saddle will put antelope, probably passing in single file at a walk, within range before they know the hunter is there.

In undisturbed country, a band with a good buck may use a very small basin with a dry pond bed as a resting place. If hills around it are high and rough, the only approach is from one of these, and from the top the range may be too long for a shot. But the band can be purposely spooked out of the basin. Then the hunter moves down within range of the dry bed, and, well camouflaged if possible, lies down in a small depression behind a low bush. There is an excellent chance that the band will drift gingerly back into the snug hideaway within a couple of hours.

A trophy buck may stay right out in the middle of a large flat where a sneak approach is absolutely impossible. A ruse that sometimes works in this case is to walk in plain sight on a shallow angle toward and past its position. While giving the illusion of simply moving on by across the plain, the hunter is actually, although slowly, closing the range, never once looking toward the buck. Very occasionally curiosity will hold the animal immobile, watching intently until, at maybe 250 yards, the hunter can drop down and get off a shot.

Earlier in the chapter, tolling antelope in by appealing to their curiosity was mentioned. This is a great sport. They also will come now and then on the run, or walking stifflegged, to the sound of a high-pitched predator call, perhaps believing it is the bleat of a youngster.

Rifles for this sport should be flat-shooting because of the long ranges. They do not need to be anything heavier than standard deer calibers. The .243 is an example of an excellent antelope rifle. It should, of course, be fitted with a scope, preferably of the variable-power type. Good binoculars are also a must item. The challenge of antelope hunting in a sporting manner is the stark simplicity of the terrain. There are few hiding places for either hunter or game. The eyesight and speed of the quarry offset any decided advantage of the reach of the rifle.

Whether one hunts the pronghorn, stalks it with a telephoto lens, or simply watches bands swirl across the rolling plains, the thrill is supreme. This is one of the most handsome of North American big-game animals; it is well managed and far from endangered. It was snatched from the brink of extinction quite literally because of the interest and determination of sportsmen. They furnished the money for the management that built back and still nurtures the national herd for everyone to enjoy. It is a conservation story of which they may well be proud.

# Bighorn Sheep
*Ovis canadensis*

# Dall Sheep
*Ovis dalli*

*Bighorn sheep are highly intolerant of human disturbance. Today they live out their lives in high, awesomely rugged wilderness where they are seen by few people. These are Rocky Mountain bighorns.*

**T**he wild sheep of North America are among the most beautiful and appealing of our larger animals. Their intelligence and wariness and the remoteness of their domains are legendary. They have played dramatic parts in the journals of early explorers, have been romanticized in many modern wildlife films, and are among the most desirable, difficult, and taxing of all trophies to acquire. Yet curiously, notwithstanding all the attention and publicity they have received, only a comparatively few Americans, chiefly hunters and their guides, and game-management people, have ever seen one alive and on its home grounds.

This is because wild sheep are not only highly intolerant of human disturbance, but also in today's world (and indeed much of history) live out their lives for the most part in high, awesomely rugged wilderness situations that are visited by only a few determined hunters, backpackers, wildlife researchers, and photographers. The wild sheep of this continent were never abundant in the way that deer, elk, and antelope were and are. Their specialized remote worlds, from the snowy ranges of northern Alaska to the steep above-timberline meadows of the Rockies and to the arid,

143

burning desert ranges of northern and peninsular Mexico, preclude authentic abundance.

Nonetheless, when the white man first knew this continent sheep were at least numerous within habitat suitable to them. The white sheep of the far north and their darker phases ranging down into northern British Columbia were never disturbed to any serious extent by man until trophy hunting for them increased substantially during this century. The bighorns of the Rockies in the contiguous states, the southern portions of western Canada, and northern Mexico, however, were hard pressed during days of early settlement and onward by meat hunters. In many places, especially throughout the southern half of their range, they were totally extirpated. The meat is delicious. Sheep helped feed many a settlement and mining camp, or went to market as a wild delicacy. Particularly in the arid southwest, where sheep were dependent upon meager water supplies, they were waylaid at waterholes and entire bands or populations were wiped out.

Competition with domestic livestock also brought the wild sheep to extinction or the verge of it in numerous locations. It is believed that the desert bighorns of western Texas' Big Bend Country, for example, received the final coup, after severe poaching, by the influx of domestic sheep from which they contracted diseases. For many years under more enlightened approaches, from the 1930s onward, there was very little legal hunting within the lower-48 states. Sheep were fully protected. Unlike deer, which have extended their ranges in many places over the past century, the history of the bighorn in particular, because it has been in such common contact with man's intrusions, is one of constant retreat.

However, reestablishment by transplants and management have been wonderfully successful. Although far from abundant, sheep are certainly not generally endangered today, thanks to the intense interest and the money of sportsmen who have pushed to make certain game managers keep whatever populations are possible on every suitable range. In fact, an idea of "where the most sheep are" can most easily be gained presently by a check of where hunting is allowed. Some of the hunts are indeed merely token, the annual cropping of a very few mature or old surplus rams. Probably Alaska has the most wild sheep. These are the white Dalls. The Yukon and the Northwest Territories also have substantial numbers, and so does British Columbia, which also has bighorns in huntable numbers, as does Alberta. Within the contiguous states, ten have bighorns in modestly huntable numbers,Washington, Oregon, Montana, Idaho, Wyoming, Nevada, Colorado, South Dakota, New Mexico, and Arizona. There are also some sheep in North Dakota, California, and Utah, with occasional token hunts currently in Utah. Some of these animals are reestablished or newly established.

Even given the comparative scarcity and the intense protection and management of wild sheep, there is still an exasperating amount of illegal killing. During the late 1960s in California, it was discovered that wealthy trophy hunters were paying high fees to a group of "guides" who took them into a refuge! Dozens of rams had been collected. In Mexico, where the desert bighorn is present here and there in remote areas in fair numbers, poaching is common. Illegal trophy shenanigans, at a price, presumably involving officials, have been going on for years.

It is doubtful that the present ranges of any of the sheep can ever be appreciably extended. Transplants do add new pockets—a mountain here, another there. But within the vast perimeters of total sheep range, local populations have always been spotty. A certain individual mountain range may contain a fair number, another none. As a good example, in the Big Bend Country of western Texas a project has been underway for years now to reestablish

## THE WILD SHEEP

**Color: Bighorn,** in the north dark gray-brown, usually but not always paler in southern desert subspecies; large conspicuous yellowish-white rump patch; muzzle, around eyes, and rear of all legs pale to whitish; short tail dark to black. **Dall sheep,** sparkling white; numerous color phases generally progressively darker from northern to southern part of range, beginning with a scattering of black hairs in tail, to a black tail, to partial black or gray saddle, to general gray with pale head and neck, to overall blue-gray to blue-black; all these darker phases with light rump patch, belly, and rear border of legs.

**Measurements, adult rams: Bighorn,** 3 to 3½ feet at shoulder, depending on latitude, the larger specimens in north; overall length 5 feet to 5 feet 10 inches. **Dall,** generally comparable in measurements to the smaller bighorns.

**Weight, adult rams: Bighorn,** from 150 to 200 pounds for desert subspecies to 300 for northern races. **Dall and subspecies,** 180 to 225 pounds.

**Horns: Bighorn,** heavy, dark brown in north to gray-brown in south; round at base, massive, tightly curled above and around ears in circle close to face; weighing as much as 20 pounds. **Dall and subspecies,** pale yellowish color; thinner, curling but also flaring rather widely outward from face; sometimes somewhat flattened toward ends; all wild sheep horns with conspicuous annual growth rings.

**Ewes:** At least one-fourth smaller; in the dark sheep much paler-colored than rams; with horns, but inconsequential, flat, slender, and not forming more than the beginning of a circle.

**General attributes:** Rounded, compact, powerful, rather short, chunky body, but exceedingly well proportioned and graceful; intelligent mien; amber to yellow black-centered eyes with phenomenal vision; unbelievably sure-footed and agile in rugged, steep terrain.

## *Range of the Wild Sheep*

Dall and Stone Sheep
Bighorn Sheep

*The desert bighorn lives in the sparse mountains of the southwestern United States and northern Mexico. It is not abundant anywhere. This ram is of record book dimensions.*

the desert bighorn, once fairly abundant there. If it ever succeeds, the final population is certain to be modest, wholly dependent upon the protection of owners of large ranches, and with only a few sheep on scattered small individual mountain ranges.

To a large extent, sheep populations around the world were even ancestrally thus grouped. Abundance was scattered. Wild sheep of wide variety and quite closely related are found clear around the northern hemisphere. It is believed that the sheep of North America anciently reached this continent from Asia. No wild sheep of the northern hemisphere is closely related to domestic varieties. But several Asian species are undoubtedly close relatives of the Alaskan Dall and its phases.

The pure-white Dall of the type species,

*Dall sheep are all white to pale yellowish, progressing to gray from the northern to the southern part of their range. This old ram with full curl horns and a heavy body is about prime for the species.*

*The Stone sheep, a Dall subspecies occupying the southern portion of the overall Dall range, varies widely in color but typically is charcoal or bluish-gray.*

*Ovis dalli dalli,* and its white subspecies of the Kenai Peninsula, *O. d. kenaiensis,* blanket much of Alaska except the western portion and reach into the Yukon and the Northwest Territories and southward into northern British Columbia. The two are considered as one in Boone & Crockett records. However, there are subtle and progressive changes in color of the type Dall toward the southern portion of the range. It is true that there may be "sports" showing darker color differences among bands of Dall sheep anywhere in their range.

This is a matter of current progressive evolution. Also progressively, the number grows as one moves south.

In the northern Yukon, for example, the sheep may appear pure white but close examination may show a scattering of black hairs on the tail. Farther north the entire tail may be black. These sheep are at the very northern fringe of the range of another subspecies, and are examples of infusions of it, the Stone sheep, *O. d. stonei,* which is typically charcoal or bluish-gray in color. But this subspecies of the white Dall varies highly in color. Some specimens appear to be perfect intergrades between white and very dark. These are gray, with neck and head gray-white. Most taxonomists list these as a distinct subspecies, the Fannin's sheep. Others, however, including the Boone & Crockett records, undoubtedly with good sense, give only one classification for color phases, the Stone. Many of the darkest Stones live in northwestern British Columbia, where the best trophy heads have been taken.

Classification of subspecies is confusing not only to the layman; scientists have long differed among themselves. One recent generally accepted authority, however, gives a listing of three phases of the Dall, including the type species, and seven for the bighorn. The Rocky Mountain bighorn, *Ovis canadensis canadensis,* is by far the most important among the bighorns. This burly brownish sheep blankets the greater share of the bighorn range, from southern British Columbia and Alberta southward through the Rockies states. Southward and spottily into the southwest United States and parts of Mexico a somewhat smaller-bodied subspecies, the desert bighorn, *O. c. nelsoni,* replaces it. The desert bighorn evolved from its spare habitat. It is by no means abundant nowadays; its existence is at least somewhat precarious, and it is managed everywhere with exceeding care.

The California bighorn is another subspecies, *O. c. californiana.* This sheep, native to California but in meager supply, is even present nowadays in southern British Columbia. There is also a subspecies called the Peninsula bighorn, of Lower California, and presumably a hybrid between it and the desert bighorn which is claimed as another subspecies. Only the Rocky Mountain type species and the desert subspecies are of importance to hunters or casual observers. The record book, always a good guide, gives a special listing to the desert sheep, but for record purposes lumps all other bighorn races together.

Wild sheep are hardy creatures, as evidenced by the curious fact that they are not found in areas where living in lush and easy, but were able to colonize not only much of the most remote and rugged country on the continent, but in extremes of climate, from the bitter winters and the far north and the high country of the Rockies to the seared and arid reaches of northern Mexico and Baja. It is logical that in such difficult living areas natural attrition is severe. Deep snows may keep bands from feeding, or cover what food is available. In the desert ranges, lack of water takes its toll.

Indeed it is believed by most researchers that the severity of the terrain and climate rather than predation are the dominant population controls. Coyotes undoubtedly kill many lambs. Mountain lions, not now abundant themselves, account for some adults as well as young. In the north, wolves do decimate the bands, drive them from lower altitudes to less congenial ranges, and probably are a definite limiting influence. Eagles are believed to kill many lambs. Throughout the enormous range of wild sheep, however, disease, parasites, and starvation, often in combination, are the chief enemies, ever waiting in the highly specialized and remote habitats outside which these awesomely shy animals are apparently unable to cope.

## Habitat

Experienced sheep hunters exploring new territory commonly judge a certain mountain as almost sure to be "a good sheep mountain." It is difficult to describe precisely what they mean. But it is true that sheep show a preference for certain peaks, and that each has its own character, and that sheep hunters and their guides probably know more about the wild sheep of North America than anyone else. Terrain varies widely from the home of the Dall to that of the desert bighorn, but certain characteristics fit rather generally for all.

The domain of the wild sheep is invariably high. To a man climbing into it, most of it is indeed rough and precipitous. Yet the specific places most sheep like best and in which they consort most are likely to be only gently angled, with broad basins and ridges with comfortable saddles scooping out undulations in them. Usually they select ridges with broad offshoots thrusting out in triangles overlooking valleys below, or overlooking an entire mountainside.

There are two nearly invariable characteristics of sheep country. A grassy alpine basin or point on a ridge must offer a broad view, for the animals depend almost wholly upon their superb vision to keep tabs on their surroundings. And there must be rough, usually steep and far more forbidding higher terrain nearby. If startled they utilize this jumbled backstop as escape territory. Almost without fail, their escape is upward. Here again the impulse is to be able to see what's going on below. They choose rough, rocky areas for escape because none of the sheep varieties seems sure of itself or feels secure running on level ground, but all are master maneuverers in difficult, rocky going.

It is true that under certain conditions of weather and season sheep use lower elevations. Desert sheep are known to cross broad, low valleys, and sheep in the north may come down to river valleys and even into timber-fringed meadows. But much of the habitat of all wild sheep is situated in the mountains chiefly above the heavy timber, and commonly where there is none. Altitudes of 8000, 9000, 10,000, even 11,000 feet or more are home at least in summer to most sheep in the far north and in the Rockies within the contiguous states.

The life of the white Dall sheep is lived out almost entirely above timberline. This is a land of ground-hugging vegetation, and of endless rock slides and steep shale slopes. At times these sheep move from peak to peak, which necessitates crossing lower valleys where higher vegetation grows. But fundamentally the highest, most remote treeless regions are their home.

It is curious that these sheep, though difficult to spot against snow, stand out conspicuously to anyone scanning snowless distant slopes from below. They seem to beckon as easy targets, but any hunter or predator who has gone after them soon knows the fallacy. Much of the best range is too rugged for use of horses. There's no way to reach the sheep country except to climb. The Stone sheep, and the intermediate Fannin's, populate a rather restricted range, and within it appear to utilize somewhat lower elevations at times than the Dall.

It should be mentioned that many wildlife hobbyists think of mountain sheep and mountain goats as inhabitants of the same terrain. While they may be found, within goat range, even on the same mountain, the goats are animals of the most awesome crags and ledges. A sheep is a surefooted animal, but it could not live where a goat can without being in constant danger of fatal accident. Goats often are found temporarily on sheep range, but sheep seldom enter the ultimate in rugged knife-edged terrain in which the mountain goat is perfectly at home.

The Rocky Mountain bighorn has a more diverse habitat than the Dall and Stone. It spends much of its time above timberline, too. But it also comes down into fringes and patches of timber. In fact, this sheep is an exceedingly adaptable animal. It must be remembered that the present range of wild sheep, and emphatically of the bighorn, is not more than a twentieth of the area covered before settlement. The bighorn does not require high mountains as an absolute, except nowadays. The first explorers and settlers in the west found them even along the Missouri River brakes, and in the badlands of the Dakotas. Scores of buttes and outlying series of rough hills surrounded by plains far from authentic

*All bighorns are skilful, assured climbers, often living on the loftiest precipices. Even lambs are astonishing in their ability to follow ewes almost anywhere.*

mountains were inhabited by sheep in Indian times. The reason almost all bighorns are in high, remote places now is that these are all the habitable undisturbed places left for them.

The desert bighorn habitat is for the most part not very high, because the mountains where they live are not as high as in the north. Much of their ancestral range was only from 3000 to 5000 feet. Many desert bighorns still live at such elevations. They are presently found along the Colorado River in southern Arizona, for example, and they often feed in Mexico way down on the flats between ranges of mountains. However, the terrain the desert bighorn selects as its favorite is possibly the roughest and most barren inhabited by North American sheep. In Mexico, for example, it seldom if ever has inhabited well-vegetated mountains easily accessible to it, but stays by choice in the most arid situations, where ragged upthrusts of rock, bleak cinder cones, and the jumbled rocky rims of northern Mexico's deserts form some of the most forbidding places on the continent.

## Feeding

On these typical desert ranges there is a wider variety of forage than one might expect, but the desert sheep must utilize more heavily than their northern cousins various types of browse plants. Wild sheep are predominantly grazers. Under proper range conditions they live on a soft diet of grasses, which account for as much as 95 percent of their intake. In the desert terrain, however, that much grass is difficult or impossible to find.

The desert sheep do eat any tough grasses available, but browse makes up a substantial part of the diet. The flowering centers of sotol are grubbed out, and buckbrush, yucca, and the green twigs and bark of the paloverde are gouged. Amazing as it seems, the bark of the viciously thorny, slender-stemmed bunches of ocotillo is eaten. Agave or century plant is in

some locations a staple. Mesquite and catclaw add to variety. Mountain mahogany and sage, when available, are eagerly eaten.

Grasses such as grama grass and needle grass are scattered on the desert slopes, and are avidly sought. One feeding habit of the desert bighorn that is notable is the large intake of varied cacti. Part of this foraging is for liquid as well as food. Rams smash the barrel cactus and gorge on the pulp. They eat the fruit and pads of prickly pear. In the relatively modest range where the huge and viciously wire-spined saguaro cactus grows, somehow the sheep are able to gouge deep holes in the trunks. No one seems quite sure how they do this without serious harm.

Farther north the Rocky Mountain bighorn has an easier time of it. Grasses are numerous and usually abundant. Vetch, needle grass, wheat grass, and fescue are common. In places sedges abound, and in others there are rushes. There is also clover, and a wide variety of forbs or weeds, most of which the sheep find palatable. In some instances where a certain plant is abundant and palatable, the bighorn is a most selective forager. It may simply snip off the budding flowers, for example, on a patch of smartweed, and leave the plants alone.

When snow arrives the diet must change. Dry grasses are still eaten if not buried too deeply beneath the snow. But now substantial amounts of browse are taken. Sagebrush and greasewood as well as rabbit brush, scrub wild cherry, willows, and alders all are eaten. If forced, the animals also snip off twigs of juniper, spruce, and fir. In spring when the aspens and birches launch buds, these furnish a large part of daily diet.

Up in the Dall sheep's domain there are varied grasses in the alpine basins, and there are also lichens and mosses to complement them. Several low-growing shrubs also furnish browse. The dwarf willow is one of these. Blueberry and raspberry bushes are also present in some areas. Invariably there are ridges swept clean of snow by winds in winter that offer good foraging locations, and of course both the white sheep and the Rocky Mountain sheep must now and then move lower to find food enough to sustain them.

Unlike the antlered animals, which routinely feed both night and day, sheep are almost wholly diurnal. It is thought that two influences have molded them thus. One is that they depend so heavily upon their exceedingly keen distance vision that they are at a great disadvantage at night. Thus they retire early and follow a routine of staying in a bed throughout the entire night. The other is that in their world they are not used to undue disturbance, and thus have little reason to be up and down around the clock.

They are, however, early risers. They are up at or before dawn, sometimes in summer as early as 4:00 a.m. to begin feeding. By the middle of the morning they are ready to rest. The younger animals are inclined to drop down just wherever they happen to be. Adults may go through a cursory pawing of the ground, pawing a few times with each front hoof on a level spot to remove stones, and then lie down. These daytime beds are invariably in a basin or in a ridge saddle or on a slope where the sheep cannot be surprised from any direction. Because most sheep are quite gregarious, and a band lies with individuals facing differently, approach of any danger is certain to be distantly spotted.

During the morning rest period the animals chew their cud. Depending upon the food supply, and the restlessness or lazy comfort of the animals, they may stay bedded until midafternoon, or they may be up having a snack at noon or early afternoon. Then a casual bed is again the rule, and more cud chewing until the serious afternoon feeding begins. These daytime beds are quite different from the night bedding places, discussed below. The late-afternoon feeding sessions in long shadows must produce bulging stomachs to last

through the night. Thus they may be long, and rather intense.

While feeding the animals are not erratic, but are thorough foragers, walking along slowly, making a business of filling up, yet always on the watch distantly for any disturbance. By the time the sun hangs low, draping the valleys in shadow and flooding a weak light across the peaks, the sheep are ready to call it a day and move toward their nighttime beds.

## Movements

This movement, which is usually brief, is sparked by an instinct and habit that is unique among antlered and horned animals of this continent. When a sheep strikes out for its nighttime bedding area, it knows exactly where it is going. The bed it will use is one that it has used many times before. If the sheep moves to a new mountain, it will make a fresh nighttime bed, but should it leave and then return to the same mountain, chances are it will use the same old bed.

The sites for these permanent beds are chosen with infinite care. As a rule the lee side of a ridge or the base of a shale slope or rocky outcrop is selected. Many are below the crest of a ridge in a protected spot. Occasionally a small cave is used. These may also be used to give protection in the desert in daytime from the hot sun in the southern range, or from rain elsewhere. The location is invariably where the approach of danger from all directions can be heard in time for an escape up into a rough area.

These night beds are pawed-out spots several feet across. From much use they are smoothed out 6 inches to a foot or more in depth. They are smelly places. Droppings are piled around the edges. When a sheep arises in the morning it generally urinates immediately, without moving outside the bed. Conceivably this is a kind of mark that stakes out the bed as belonging to a certain individual.

Because sheep country is high or arid, with little moisture in the air, the beds seldom become soggy but remain dry. Some of these beds show evidence of use for several years.

The hooves of sheep are fashioned with sharp, hard edges and with a concave interior filled with spongy tissue. Although sheep seldom walk the knife-edged pinnacles and crags that goats casually traverse, they are fantastically sure-footed and agile. Rams when disturbed often plunge nonchalantly down a 50-degree slope. They are able to walk down nearly vertical cliffs if there are even the slightest small toeholds en route. When it is necessary to jump, the animals have no trouble covering 15 feet or more across a chasm, and easily fly as high up as 4 feet in the air on the way.

Jimmy McLucas, who for many years trapped big game for transplant for the Montana game department, once kept losing rams from out of an enclosure where they were being temporarily held. It was surrounded by a plank fence 8 feet high, the planks 2 inches thick. He caught a ram in the act of leaping, touching the 2-inch top of the fence with all four feet and bounding down to the other side.

Most of the movements of sheep are a casual walk. If disturbed they may trot for some distance, and when severely frightened they gallop up a slope at high speed. On level ground they may be capable of 25 to 35 miles an hour, but only briefly. Even when undisturbed they are masters at leaping across broad fissures. Or they casually jump off a ledge to land 20 to 30 feet below. Their sturdy legs flex like springs to bear the body weight as they land on all fours.

A few sheep hunters and naturalists have observed a most interesting and amazing ability of sheep in ascending extremely broken country. A vertical cleft in a cliff may rise from its base to a mesa above. It may be several feet wide, with minor horizontal ledges a few inches wide here and there on either face. Such formations, often used, sometimes have been called "sheep ladders." They are like chimneys open on one side, in the face of a cliff. Fred Bear, the renowned archer and wildlife photographer, once saw a group of sheep go up such a ladder. Lining up single file, the first one leaps upward, strikes its hooves against one side, aiming for the slightest hold and sometimes none, then flips aside in midair to bound again, higher, and strike the other side. With a few quick left-side, right-side leaps, pushing each time outward and up, it tops out, and the others follow in their turn.

Although wild sheep cannot be termed truly migratory, many of them do move seasonally. These movements are entirely dependent upon weather and availability of food. Cold does not bother them. Wild sheep do not bear wool like domestic sheep. Their coats are of hollow hair, which provides superb insulation. But no matter how well shielded they may be against bitter weather, if snow gets too deep for them to move about, or covers forage deeply, they must move down to a lower altitude.

It is common for some sheep bands to winter at only 2000 to 4000 feet, far below their summer range, and sometimes requiring a trek of 25 miles or more. Some, however, are able to stay all year on the same range. The northern sheep seldom have any difficulty finding water, but the desert bighorn sometimes must travel far to a remote spring. In summer a movement to an area of reliable water may be necessary. Desert sheep do not necessarily go to water daily. Sometimes they manage several days without drinking, utilizing cactus pulp to satisfy their thirst.

All sheep are to some extent wanderers. They may move whimsically from slope to slope or mountain to mountain within their domain. Conversely, some of them stay for weeks within a very small range. The white sheep, most observers believe, are less inclined to wander than the bighorns. All sheep avidly seek salt licks. These may be at a salty spring, or formed of clay inlaid in a bluff. Occasionally the animals find brittle rock deposits that contain salt, and chew these. Mineral licks are constantly visited. Some show signs of use for many years and even are responsible, researchers believe, for keeping sheep in the area. Studies have shown that trips to a particular lick sometimes may cover 10 or 15 miles.

## Breeding

There is a definite caste system among sheep during most of the year. The rams form small groups, and these are usually composed of males of similar age. Here and there an old ram is a loner, perhaps crotchety and unsociable, or else off by himself because other

*Combat among rams for the right to breed during the annual late-fall rut is spectacular to see. After posturing and maneuvering, on some signal rams rear onto hind legs and then lunge forward to smash head to head. The impact is loud and ringing and carries for great distances. Rams of nearly equal strength continue these tests for dominance for long periods, until one concedes.*

more vigorous rams bedevil him. The rams usually spend the summer higher up than the ewes and lambs. But by late fall they cease being quite so friendly with one another and begin to seek out the ewes.

As early as October in the north, rams begin to play at fighting, locking horns, and wrestling, or shoving. But presently the play becomes rougher, and finally there are awesome battles. Curiously, rams do not fight for the favors of one or several ewes. They do not attempt to form harems, although one may try to dominate several ewes briefly. Actually the rams are wholly promiscuous. Two may battle fiercely, then one backs off and the other breeds a nearby ewe. He may then leave her, seeking another, while the ram he was battling with breeds the same ewe. The ewes are in heat very briefly. And the rams have the unique physical ability to breed several within only a few minutes.

When a battle builds up between two rams, they may approach each other as if unaware. The routine is a highly stylized ritual. Suddenly when a short distance apart, and not even necessarily looking at each other, both animals rear up on their hind legs and rush together. They may drop down again and then rush headlong, but usually as they strike head-on both have the forefeet off the ground, slamming with full body force and momentum. The massive horns came together evenly; the left of one is matched with the right of the other. Observers have heard the crashing sound as much as a half-mile away.

Immediately after each impact, the rams rear their heads back and back away, often shaking their heads. Then as if on signal, they smash together again. They appear to gauge very carefully the head-on strike, to avoid severe physical damage. Occasionally severe injuries do occur—horn or eye or brain damage. The length of the battle depends on how evenly matched the rams are. Eventually one will break off. The winner may quickly breed a nearby ewe several times, then wander off seeking other conquests. No ram "owns" a ewe.

The peak of the rut in the north is in late November and into December. This assures that the lambs will not be born until early spring has come to the wintering grounds. Among desert bighorns, however, breeding is much earlier, and the rutting period lasts much longer. This is a device of nature to assure that the lambs will be born at a time when winter rains have brought green and abundant forage briefly to the desert. And the long season, which may start as early as August or even late July and run on through into fall, helps to assure that the desert sheep, spread thinner in population than their northern relatives, all will find each other and mate.

## Birth and Development

Once the rut is finished, the rams revert to their male-chauvinistic ways, disregarding the ewes entirely and joining in groups of their own once more. The sexes may be forced together on a winter range. How closely depends on the quality of the range. Six months from the time the ewes were bred, the lambs are dropped. This places lambing time, except for the desert sheep, in May and June. A single lamb is the rule, although there may be twins.

When it is time for her lamb to be born, each ewe leaves her group and seeks a high ledge or the foot of a cliff high up, where there is protection from weather and from where she may keep watch over a broad area for danger. Lambs of the bighorn are a fuzzy dark gray with a darker streak along the back. Those of the Dall are white. Both stand less than a foot high at the shoulder and weigh about 8 pounds. Rocky Mountain bighorn lambs may be a bit larger.

The ewe is a nervous, constant guardian. The lamb is able to stand on wobbly legs and nurse within a few hours, and it grows swiftly.

*This bighorn lamb is six to seven months old. Born in May or June, and weaned months before, this lamb has probably spent its entire life with its mother and other ewes and lambs. Wolves are this youngster's most serious predator.*

the fall. Although wild sheep are silent most of the time, the young lambs do bleat, and the ewes now and then answer with a blatt. Even an adult ram may blatt explosively in challenging another. But by and large sheep are not vocal creatures.

By the time the lambs are weaned their horns are sprouting. Yearlings and even two-year-olds still stay with the flock, and particularly the lambs of the year and the yearlings play together, but ewes become annoyed if the large ones get too rough, and will run them off. Occasionally a single ewe baby-sits with a group of lambs while the others feed nearby. The mixed flock contains no mature rams, and indeed adult rams are utterly disdainful of the lambs. Curiously, however, when forced into contact with young and ewes, adult rams are often stoic and solicitous. Big-game trappers transplanting bighorns have discovered that when moving sheep by truck, for example, the animals are quite calm, intelligent, and sensible. A big ram will allow a lamb to crowd under its belly and will not harm it. Conversely, goats under similar circumstances sometimes will kill each other.

The long two-year-old rams are not allowed with the bachelor gangs of mature rams. Some wander away from the flock in their own groups and some do not. It is believed that the first successful breeding of the rams occurs when they are long three-year-olds. Ewes are able to breed when a year younger.

At breeding age the horns of the rams have

But as a rule the ewe keeps it at the birth site for about a week, leaving only to grab a quick bite of forage now and then. Often while the younger sleeps, the mother stands alert guard. After seven or eight days the lamb is allowed to follow closely with its mother. Presently she rejoins the band of other ewes and their lambs. This band also contains young of the previous year, still hanging with the ewes. Lambs of both ages play vigorously.

Soon the young are nibbling at grasses, and within a month most of them are at least partially weaned. During this training period the mother lets them start to nurse, but then abruptly leaves. Here and there a less firm mother allows a lamb to continue nursing on into grown to roughly a half-circle. Hunting regulations in most places require that a legal ram have at least a three-quarter curl. Such animals are in the four-to-five-year age class. Seldom is a full curl attained until a ram is at least seven. At twelve to fourteen a ram is very old, and some of the best trophy horns—age can be deduced from horn growth rings—are from rams in this age group. A few animals may live longer. A young desert ram of known age brought to the Black Gap Wildlife Management Area in the lower Big Bend Country of west Texas in an attempt to reestablish desert sheep there lived to be seventeen.

Because of the massiveness and beauty of wild sheep horns, they have been a fetish of sportsmen throughout the history of hunting all around the world. Horns of the bighorn sheep generally curl in a wide C around behind the ear and in a circle forward back around toward the massive base. Most, but not all, grow rather close to the face. Because full-curl bighorn horns are so heavy, they often interfere with an animal's vision. The ram then rubs and scrapes the ends against rocks and "brooms" them off. Many of the finest trophies have broomed horns. Some, of course, may have the tips broken during battles, but among the bighorns brooming is common.

The white sheep and their relatives have a somewhat different horn configuration. They are not so massive, are yellowish instead of

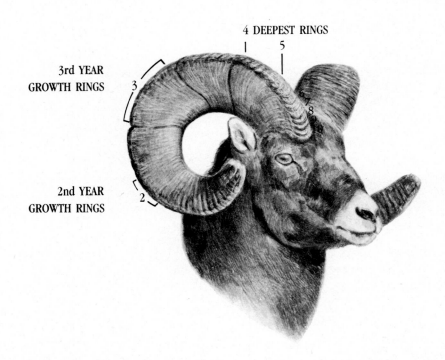

*A ten-year-old bighorn ram with a superb set of horns. The growth rings, starting from the tips, clearly show the animal's age. The tips have been "broomed" by the ram, else they would have obstructed his vision as the horns continued to grow.*

*Male bighorns have heavy, brownish, tightly curled horns that describe a circle close to the face. A large ram's horns may weigh 20 pounds or more. Generally, the older the ram, the more massive the horns.*

brown in color, and almost always flare widely out away from the face. They are thus broomed less, because they are not so likely to interfere with vision. Numerous specimens of Dalls and Stones have more than a full curl with perfect tips for that reason, and also a wider tip-to-tip spread because of the flare. The Stone sheep generally bears horns slightly more massive than the Dall.

To illustrate with measurements the basic characteristics of sheep horns, the longest bighorns in the record book are presently: right, 49½; left, 48¼. The tip-to-tip spread is 23⅞. The first-place (and longest) Stone sheep horns, the longest sheep horns taken to date on the continent are: right, 50⅛; left, 51⅝; tip-to-tip spread, 31 inches. The top-ranked Dall: right, 48⅝; left, 47⅞; tip-to-tip spread: 34⅜. Horns of the desert rams, even though the animals are often smaller in body, rank in length, spread, and base circumference just about as large on the average as those of the Rocky Mountain sheep. Comparative base circumferences of first-place records give a good idea also of horn types among the several varieties. Rocky mountain bighorn: both horns, 16⅝. Desert bighorn: right, 16¾; left, 17. Stone: both horns, 14¾. Dall: right, 14⅝; left, 14¾.

*This ram with heavy horns has survived another winter and now is shedding and very thin. It may recover fat and strength during the summer, but its powers are waning.*

## Senses

Like that of the pronghorn on the plains, the visual ability of wild sheep has often been likened to that of man assisted by an 8-power binocular. Sight is far and away the most important sense. The immense vistas among which the sheep live allow them to scan incessantly and over great distances. In somewhat the same way as an antelope of the plains, they also keep danger at a distance by recognizing it from afar. Unlike antelope, however, sheep cannot escape predators by outrunning them on flat expanses. Buy any predator has

a difficult time apprehending an adult sheep, first because it sees the danger long before it is near, and second because when it takes to its heels in the broken high terrain always nearby, leaping off ledges and bounding across gashes in the rocks, it is practically uncatchable.

Sheep do not always recognize immobile objects. A prone hunter glassing from a distant slope may be passed over with a cursory scanning. But the slightest movement brings the sheep alert and curious. In fact, curiosity now and then gets a ram into trouble. A pair of hunters, knowing they have been sighted because a bedded band has arisen and stared at them, sometimes successfully use the ruse of having one stay put while another drops off a ridge behind the circles for a stalk.

Interestingly, when sheep spot an intruder, it seldom works to try to get out of sight quickly. A hunter, knowing he has been seen distantly, has an urge to duck back over a nearby ridge crest. The moment he disappears, every sheep usually will run. Further, if a sheep is spooked so badly it gallops away, almost without fail any sheep in the vicinity will race away. Sheep do not often run at full speed. When one does, the sound alerts all others that this is no time to be wondering what is wrong.

Scenting ability is well developed, but it is not always very useful except at close range. Air currents and winds in the high country are whimsical. If sheep do catch a scent they instantly give evidence of it in alertness or flight. But they seem to know, by the way they select places from which they can see over a large area, that every-which-way breezes may not be dependable.

Hearing is also keen. But again, it is only moderately useful. As noted, the sound of other running sheep tells of danger. But a falling rock or a rock slide means little. Mountain country is full of such natural sounds. The unnatural sound is something else. A distant human voice, the rhythmic sounds of a packtrain moving in rocky country, the blowing of horses, the clang of any packtrain or hunter equipment, all are frightening.

## Sign

Although hunting guides are not uninterested in sheep sign, in most situations the expanse of country is so vast that it is far easier and more profitable to look with powerful binoculars for the sheep themselves. Because of their wandering habits, tracks and droppings and beds may not mean there are sheep present on that mountain. Tracks of desert

sheep at a waterhole are, of course, an extremely important clue. Tracks and gougings at a mineral lick also tell that sheep regularly use it.

Sheep tracks might be confused on some ranges with those of deer. In mud the sheep may leave a depression in the center of the track because of the spongy center in the hoof. Deer do not. The general outline of sheep tracks when prints are clear, however, show nearly straight edges, and they are blockier, larger, and less pointed than deer tracks, and almost as broad at front as at rear. In front the toes, when not splayed out as in soft earth, are more prominently separated than those of deer. In a few instances goats may be on the same range. This might be confusing. Goat tracks if plainly printed tend to be more square, and a bit indented along the middle of the outside edges.

Sheep droppings can be an indication of population, and current resident if they are fresh. They are not at all easy to distinguish

*A three-month-old Dall lamb.*

*In summer Dall sheep are often easy to spot high up on green mountain sides. Some of the most remote and untracked alpine lands of the Rocky Mountains are the final sanctuaries of native wild sheep, that serve as perfect symbols for the rapidly vanishing wilderness.*

*Mountain goats are well equipped to negotiate the precarious crags and slopes of their habitat. Their convex hoof pads form spongy cushions that grip the rock.*

from either deer or mountain goat droppings. Deer droppings as a rule are rounded on both ends while those of sheep are bell-shaped. However, droppings of deer take several forms, and those of goats are often as bell-shaped as sheep droppings. Like all of the ruminants, sheep droppings form masses or elongated shapes when forage is soft, and show as hard pellets when food is less so. Observers have often found droppings made up almost wholly of clay after sheep had been using a mineral lick.

Undoubtedly nighttime sheep beds, discussed earlier, are the most reliable sign. One who recognizes these cannot possibly confuse them with signs of any other animal.

## Hunting

Successful sheep hunting is an art. Like many others, it is deceptively simple. There is no wide variety of techniques and methods, as there is in deer hunting. For example, you don't spook a ram out of a canyon by pitching rocks into it, as sometimes works for deer. If you are close enough to pitch a rock, any ram that had been there long ago counted the whiskers on your chin and left.

Put as simply as possible, the technique of successfully taking a sheep is to see it before it sees you, study it to make sure it is a legal specimen and a trophy of proportions you will settle for, and then get into shooting range without being seen. Any sheep that has spotted a hunter, even at a couple of miles distant, probably cannot be come upon easily within range unaware.

In any sheep terrain, from Alaska to Mexico, the hunter must be endowed first with stamina, and then with patience. Possibly one is as important as the other. Without both, success is unlikely. In almost all instances, a hunter is going to have to climb, in tough country, and the stalk may require hours. Mandatory equipment, along with proper clothing and boots, is: quality binoculars, preferably of about 9-power; a spotting scope of at least 20-

*These Dall sheep are resting in their daytime beds, each animal facing in a different direction to watch for danger. The vision of wild sheep has been compared to that of a man looking through an 8-power to 10-power spotting scope. The sheeps' ability to see approaching danger from great distances has been a key factor in their survival.*

power with a small tripod attached; a second flat-shooting rifle in a class from the .30/06 to the 7mm magnum capable when necessary of long, lethal shots.

The basic routine for sheep hunters is to select a mountain or an expanse of country that seems suitable, then ease into it inconspicuously and sit down—never stand or sit skylighted—in a place where a view of several slopes and basins is possible. The sitting spot should be where the hunter is not obvious. Then the glassing begins. Glassing for sheep doesn't mean just sweeping binoculars across the region. It means first making a slow sweep of all of it in view with the naked eye, and then a yard-by-yard study of it with the glass.

The white sheep are rather easy to locate most of the time, and the gray-faced Stones are not too difficult, but the bighorns are likely to blend well and be tough to pinpoint. The hunter using a glass may see some minute distant spot that looks interesting. He keeps coming back to it. Perhaps after half an hour he catches a movement. Or he is interested enough to get out the spotting scope. The spotting scope can bring in information about the sex and size of the sheep. A lone sheep is very likely to be a large old ram, and a most desirably trophy. However, groups of rams, three to ten, all of the same general age class, may all be trophies, and then a most careful and anguishing decision must be made.

An experienced sheep hunter does not begin planning his stalk the moment he spots a trophy. Unless the animals are close, and a stalk is easy and fast, going after feeding sheep is a risky business. After a long, exhausting stalk, the animals may be long gone, and a mile away. By knowing the feeding and daytime bedding habits of sheep, one can judge from the time they are sighted about how long it will be before they lie down, or, if they're lying down, about when they'll be getting up to feed again. Sheep spotted, let's say, at 8:00 a.m. probably will lie down soon, at least within the next hour, and then stay bedded until at least noon. Sheep seen feeding at 1:00 p.m. will probably lie down shortly for several hours, but then will be up feeding and moving late in the afternoon.

Most hunters hesitate to try a stalk on feed-

*Penetrating into the best sheep range means making plenty of footprints over rugged terrain. Hunting is rarely easy and requires a lot of glassing. Bighorn rams have telescopic vision and an uncanny ability to escape over open ridges at the approach of humans or natural predators.*

ing sheep late in the day. It probably will not be successful, and it may be a long way down for the hunter in the dark. However, the animals should be watched patiently. If their bedding location for the night is ascertained, or a guess made by the look of the terrain that it is nearby, they will probably be up and feeding right here the next dawn. Sometimes rams will select a basin in which to lie down in the daytime where there is absolutely no cover from any direction to bring one within possible range. Experienced hunters in such circumstances simply resign themselves. They keep a watch on the sheep until they finally move. Perhaps then a stalk may become possible. If not, it is better to know the general location of the animals and try again tomorrow.

When forage is abundant, on many ranges sheep do not take a noon snack, even though they usually do. If not, they may stay bedded for four or five hours. Once a hunter is certain the sheep have not seem him or are not studying him with concentration, it is simply a matter of making the stalk without being seen. "Simply" probably is not the word. Once in a dozen tries the terrain may lie just right for an easy stalk. Mostly, however, no sheep stalk is easy.

The plan should always be if at all possible to circle and get above the sheep. They do not seem to expect danger from above and do not look up often. They lie overlooking a vast expanse below. Coming in from above and behind offers the hunter immense advantage, provided he does not have problems with wind directions. Even though sheep may not use their noses as much as their eyes, it's dead certain that if one winds you it will flee.

If a stalk is long, the first part can be covered swiftly, given ample cover. But the last stretch should be done with infinite patience and with cautious and thorough pre-glassing of the immediate area. Sometimes sheep get up and move a bit. They may have moved enough so in haste you stumble into them where you don't expect them. Or there may

have been an animal or two unseen before the stalk began. You bumble into one of these and spook all the rest.

Practically all modern-day sheep hunters are guided. Thus they do not need to know very much except what the guide tells them. But to be successful they must pay close attention, and realize what sort of sport this is. No two stalks are alike. Meticulous plans for each must be made. Fundamentally the idea is that with a proper stalk the shot itself should be easy, at an immobile animal, and if possible not over 250 yards at most. It is the stalk that sparks the excitement and challenge. On many stalks a guide and hunter spend seven or eight hours from the time the sheep are first sighted until the shot is made.

Knowing what is a trophy and what is not is extremely important. Most guides are expert judges. It is sometimes very tricky to judge whether a head is a good or poor trophy. Some young rams among the white sheep, for example, may exhibit a full curl but still not be much. A curl that measures in length in the low 30s would never be accepted by a true trophy hunter. But one of 38 to 40 inches would unquestionably be most desirable.

Overall, sheep hunting is a physically difficult but fascinating sport. Part of the appeal lies in the wild country that is the domain of the animals. Part of it relates to the fact that, even though sheep are not rare, the number of permits for any sheep variety nowadays is limited. One may apply in several states of the contiguous United States every season for years before drawing lucky—or maybe never making it. This, too, heightens the drama. Whether one hunts, or simply goes into sheep country to see or photograph the animals, it is a good feeling to know that careful scientific management, the many attempts at reestablishment on denuded ancestral ranges, and the transplants to new ranges have been successful over the years and that without question the wild sheep will be available in at least limited numbers far into the future.

# Mountain Goat

*Oreamnos americanus*

The mountain goat, sometimes called the Rocky Mountain goat, is a collection of contradictions. It isn't a goat, but rather a very distant relative of several totally dissimilar animals of Asia and Europe that appear to be a link between the goats and the antelopes. It has been on this continent, scientists believe, for well over a half-million years, and yet less is known of its life history than of those of our other horned and antlered creatures.

It appears at casual observation to be a stodgy, clumsy creature, slow and ungraceful in its movements. Yet it lives in the highest, steepest, most dangerous terrain North America offers, calmly staring down without concern from narrow ledges on which it cannot possibly turn around, perhaps over a cliff falling perpendicularly 1000 feet or more to the next ledge. Its life is spent at the top of this continent's world, in climatic conditions under which man can survive only with most careful planning and outfitting and then only for a few days at a time. Yet the goat stoically endures day-to-day bitter weather and violent storms, and grows rolling fat on what would seem to be a meager food supply.

The mountain goat is wholly American.

*Male goats live at elevations frequently battered by storms, where the vegetation is sparse, and where snowfields may linger all summer. The atmosphere is thin and cold. The ground may resemble the brittle surface of the moon. But the goats survive well there.*

There is no other similar animal in North America, and none elsewhere. Scientists believe that the ancient forebears of the mountain goat came to this continent from Asia when there was, they suspect, a land bridge connecting the continents across the Bering Strait. Presumably the distant relatives of the mountain goat include such curious creatures as the goat-antelopes of Asia, the chirus, the curious goral of Siberia, the cliff donkey or serow of western China, and the better-known several varieties of chamois.

None of these is remotely similar to the mountain goat in physical appearance or clothing. Presumably the goat, isolated in its far-northern North American home, slowly evolved to fit the terrain. The white coat may have been designed by nature as camouflage in snow, yet the animal is commonly seen in snowless areas among bare rocks or even occasionally dark spruce, where of course it is blatantly obvious. There is really small need for contriving camouflage. In the world of this creature, enemies except for the terrain itself are few, and visits by man are brief and infrequent.

It is interesting that the mountain goat apparently has never been numerous—that is, abundant in the sense that other hoofed animals such as deer, pronghorn, and buffalo have been. But it also has never been scarce or rare within its range. Observations of early explorers, and modern studies and observations over many past years, tend to prove that the goat population has long remained at a modest but unendangered level, with only moderate periodic fluctuations, seldom of consequence. Possibly the awesomely forbidding habitat of the creature acts both as a protection, by eliminating or controlling populations of potential enemies, and also as a check on overpopulation by the animal itself.

Scientists admit that aside from occasional transplants of goats to new and suitable territory to launch new bands, the mountain goat is for the most part beyond the reach of modern game management. This is because its living area is also too far removed, vertically, from the common reach of man. And so it may be that this creature is one of the few and classic examples of a near-perfect balance in nature. Further, because man's influence is not expected to touch the realm of the mountain goat more than lightly in foreseeable times, this may be the one creature native to North America secure forever against possible listing as rare or endangered.

Although sport hunting has never harmed any North American animal population—under modern management techniques, indeed, it is utilized as a tool for keeping population levels tailored to available range—hunting has had and presently does have no influence whatever on the goat population. There are several reasons for this. Most important probably is that of all native big-game animals, the mountain goat is the least desirable as a trophy. Hunters after sheep may decide to try for a goat as an incidental, or to fill out a collection of trophies. However, there is so little difference in horn length among mature goats that the chances of taking an exceptional head are remote. "Exceptional" doesn't mean much—possibly another inch of horn above the norm. Further, the difficulty of getting to goat country, accomplished only by a tough and often dangerous climb, puts the animal out of range of a majority. The meat is edible, but most find it far from desirable, and useful only in a pinch.

Little use has ever been found for the hair of the mountain goat, and even if an important use were discovered, there are not enough goats to make it worthwhile, aside from the difficulty of collection. Northwestern Indians long ago made yarn from the fine undercoat of the animal by rolling it into strands. This yarn, sometimes dyed, lent itself to weaving. The source for the wool was usually animals that had died naturally, or from areas where it had been raked off by brush or rocks during spring shedding.

Another reason that the goat is of little in-

## MOUNTAIN GOAT

**Color:** White, often shaded with yellow; eyes, horns, nose, and hooves black.

**Measurements, mature billies:** 3 feet to 3 feet 4 inches at shoulder; overall length 5 feet to as much as 6 feet.

**Weight, mature billies:** To 300 pounds.

**Horns:** Sharp, slender, cylindrical, stiletto-like, backward-curving, at maximum to 12 inches in length, generally 9 to 10 inches; permanent; grow throughout lifetime, with annual growth rings.

**Nannies:** Somewhat smaller even at maximum; horns seldom over 9 inches, but one record-book specimen taken over a half-century ago measured over 12 inches.

**General attributes:** Hair long, shaggy except on lower leg; humped shoulders; both sexes with beard on lower jaw; blocky overall appearance; calm, deliberate temperament; exceedingly surefooted.

*Range of the Mountain Goat*

terest to hunters is that it is really not a very challenging quarry. This does not mean it is not an interesting animal—simply that it is not especially alert, wary, or imaginative. In personality the goat is a plodder, stolid, a kind of stubborn ascetic. One writer has suggested that its lack of imagination and wariness is undoubtedly one of its greatest assets. If it spent time dwelling on the precariousness of its perches, he claims, it would quickly be a nervous wreck. Further, its plodding nature is unquestionably a by-product of its chosen habitat. Any creature in a hurry here would not last long.

No one knows just why the mountain goat has never come down from its above-timberline pinnacles and crags to probe into other areas. But it is certain that this love of the high places has allowed the animal little leeway for colonization. Apparently it has never been able or physically equipped to push its range downward. Thus only in places where it was able to move from one mountain range to another along ridges leading from one tall peak to another was it capable of enlarging its domain.

So far as is known, the original range was from Alaska southward as far as a part of the Cascades in Washington, and along the mountain spines running into central Idaho and western Montana. Over the years transplants have been made, often with great physical difficulties for the game-department personnel doing the work, from range to range within states where the animals were native. This has been heavily emphasized in Montana.

Transplants have also been made to suitable territory in states where goats were not native. Today the animals are found in their original ranges in Alaska, the Yukon, the Northwest Territories, British Columbia, western Alberta, Washington, Idaho, and Montana, plus in areas of the last three to which they have been transplanted. Alaska undoubtedly has the largest population. Within the rather restricted range in the lower-48 states, Washington contains the highest number, interestingly, possibly half as many as are in Alaska. Transplants to new territory within the United States have been made to Oregon; that project was begun in the 1950s. There is a small but stable population in northwestern Wyoming, a very few placed in mountains east of Salt Lake City, Utah, and a quite substantial population in western Colorado, heaviest in the Mount Evans and Collegiate Range areas.

One of the most interesting islands of goat population is in the Black Hills of South Dakota. The basis of this herd was formed by a few escapees from a small Canadian group placed in Custer State Park over half a century ago. Once free, they moved as if by compass reading straight for nearby 7,242-foot Harney Peak, the highest point in South Dakota, and to the adjacent Needles—the only two high, rough areas in the Black Hills. This herd well illustrated how the highly specialized goat is tied to high, rugged terrain. The animals over all the years have never spread outward from those two colonizing points. Populations have built up as high as 400 animals, but undoubtedly cannot ever exceed that because of lack of living room and food.

Certainly there are still left a number of places in which the mountain goat could survive and propagate, perhaps on peaks in northern New Mexico, California, or elsewhere. But the likelihood of broadening the range of the animal is slim. It cannot reach new ranges except with man's help—that is, by game-management people from the states involved. Because the goat is not avidly desired by numerous hunters, and because management personnel have more than they can handle with the more popular and abundant game animals, there is little interest in further transplants.

The goat population of the present, however, is in a healthy condition that will continue. Enemies, except for weather, snow slides, possibility of fatal falls (which do occur), and minor parasitic infestations, are minimal. Predation is low. Eagles are thought to kill a few kids. But in the sparse domain of the goat, few other predators can survive. On occasion goats are forced by blizzards to move down into or to cross high valleys. Here a wolf, mountain lion, bobcat, or coyote may attempt a kill. A few do succeed, but the goat with its rapier horns is a tenacious and dangerous foe, not inclined to give up easily.

Scientists have made some attempts to classify goats into a number of subspecies based on geography. There is little reason or need to do so. As a rule mountain goats of the more northerly range are larger at maturity than others. A few most unusual male specimens have scaled 400 to 500 pounds. There are minor differences among specimens from different latitudes and mountain ranges. But all are so similar that subspecies classification seems at best a questionable or unnecessary practice.

## Habitat

"Mountain" is certainly the correct qualification for the favored and indeed mandatory habitat of this animal. But that word doesn't say it all. It is true the goat does appear at times below timberline. But its classic preference is for the treeless country, where only dwarf scrub appears, and for the barren rocks and crags and short-grass mountain meadows above the line where true timber grows.

It spends much time on completely barren slopes of jumbled rock. At the bases of desolate crags there may be scattered, tough high-altitude plants subsisting in stunted fashion in the thin, flinty soil. But the rocks replace, for cover, what the forests offer to animals lower down. Portions of each day may be bright with sun, but arctic winds incessantly whistle and cut, and warmth is little known even when the thin atmosphere is cloudless.

The goat is oblivious. Its shaggy outer coat of hair may be as much as 6 or 7 inches long. It is nearly waterproof, or at least water-repellent, coated with lanolin and thus similar in "feel" to hair coated with beeswax. This outer coat is a buffer against wind and snow, and also against the dampness eternally present in much of the animal's range, such as the steep slopes along the Pacific where fog and rain are normal, not unusual, conditions. The undercoat is composed of extremely fine wool, and may be 4 inches deep. This protection is a means of adaptation to the severe habitat. The coat is gradually shed, beginning early in the brief summer, but the animals are never without partial protection.

It is true that hunters often hunt moose, sheep, caribou, and goats all on the same trip, with a bear also thrown in occasionally. But typically the terrain goats prefer is far more rugged and broken, and also higher, than that inhabited even by mountain sheep. The sheep, for example, commonly move downward during severe winters to foothill country. A group of goats may take up residence temporarily in highest timber edges or protected high valleys, but for the most part they confine themselves to the steepest talus slopes and rocky ledges. It might seem that nothing could grub sustenance from the crevices and between the rocks, but goats put on an amazing layer of fat each season, and seem not to have the slightest concern over the severity of their surroundings.

The goat, in fact, will go into high places where no sheep would venture. It is common for an old billy to stand for hours skylined atop the highest, most precarious crag in the entire vast expanse of its bailiwick. One may lie on a narrow ledge, forefeet and head hanging over the edge as it sleeps. Or it may stand or lie down atop a needle spire of rock that appears to afford no way up or down, or even room to lie at the top.

It is most interesting to realize that these

*In summer, goats shed the long white hair of winter to grow a new coat. So for a time they are thin and shabby looking. By autumn the hair is long and flowing again, hanging to their black hooves.*

animals actually enjoy the terrain, will not exchange it for any other, and, far from suffering in it, thrive and are placid and content. Of course, physical evolution and adaptation give them confidence. The foot, for example, is specifically adapted to the needs of living where climbing or descending very nearly perpendicular, sheer rock faces may be an everyday occurrence. The hind-foot hoof is a bit smaller than the front. Both are broad, and each hoof very nearly forms a square. Inside the hard outer portion the pad thrusts outward in convexity, rather than showing the concave form of most hoofed animals. Those spongy cushions grip the rock with astonishing traction. This enables the goat to climb and descend the steepest of cliffs without difficulty.

It was mentioned earlier that the mountain goat does not seem to be a very alert or wary creature, that it is a stoic perfectly tailored to the continent's most bleak and rugged habitats. It may be that its stoicism, its fundamental lack of fear or skittishness, has evolved because it lives in a world where nothing much happens, or at least nothing that can be avoided. It never knows the danger of the avalanche—a fairly common cause of goat demise—until one strikes; it fears no fall until it falls; it is seldom shy or worrisome of enemies because so few ever appear. Other earthbound creatures, including man, may consider the habitat of the mountain goat a place fraught with dangers. The goat, born here, and undoubtedly of only moderate intelligence on the animal scale, must consider this upended, windswept, glacial terrain the safest place on earth.

## Feeding

The high peaks are not only pleasant and reasonably safe for this highly specialized animal, they also offer a much easier foraging life than one would believe. To be sure, the choice of food at any given time is not very broad. But there is enough of it. Hunters have often marveled at the fact that mountain goats can make a living in the high places which seem so barren, where any timber is dwarfed and sculpted into agonized shapes by the incessant winds. This is only because the human intruder does not look closely enough.

Most abundant of the basic items of food are the mosses and lichens. These grow abundantly well above the timber. They may be short and clinging in rock crevices. But the domain is vast, and the goat can roam as widely as necessary, picking here and there. Like the other horned and antlered game animals, the mountain goat is a ruminant with a multiple-chambered stomach. It feeds for a time, then lies down to chew its cud.

In addition to the lichens and mosses, which are utilized around the year, there are dwarf willows, taken as browse. Browse makes up a substantial portion of the diet, particularly because summer is short and low browse plants are available when green succulents are not. The varied small, tough willows are a staple browse item. Added to these are dwarfed birches and aspens where present, and blueberry and bearberry brush. Occasionally some juniper is available, and when the animals move downslope a bit they are able to find pockets of high-altitude conifers, some of these also dwarfed by altitude.

So far as is known, goats are daytime creatures. They are not known to feed at night. This probably indicates that they are under no urgency to do so, and speaks well of their adaptation to what would seem to be a rather sparse food supply. It also undoubtedly indicates that they never have been harassed to any extent by enemies. They have no fear of moving about feeding in daylight. They are up and browsing placidly well before the sun, and continue to midmorning or after.

Nothing hurries this animal. When it sets out to eat, it potters along, pauses to look off down into a valley or at another ridge, consorts casually with others of its kind unless it happens to be a loner, pokes about here and there grabbing a bite, moving a step or two and ferreting out whatever is handy. If the wind is a gale, it doesn't care. If snow swirls, it continues to feed as if unaware. But goats don't like rain. They may be forced to accept it at times. Usually, however, they will retire to some cave or overhang to wait until the shower passes.

Sometimes in winter they will seek windswept areas purposely, to find forage where snow has been whisked away. Or they may go to somewhat lower elevations where the conifers and aspens furnish food. This seldom is a long trek and can hardly be considered a migration to a winter feeding range. In the short spring and summer, grass explodes from the edges of the snowbanks. On slopes where winter slides have swept brush and trees clean, grass takes over with the spring, or at least by the second summer. These patches are important locations for feeding.

Well over half of the summer rations are composed of grasses and alpine weeds or forbs. The greenery of the summer-leaved shurbs such as bearberry and willow add quantities of semi-soft foods. If aspen is present it is much sought, summer and winter. In live-trapping operations in Montana for transplants, one trapper for some years used aspen most successfully as a bait in his corral-type traps.

Most of the middle of the day is spent resting, sleeping in a cool spot, which in summer may be a patch of ice or snow, or a high crag where the cool wind whistles, or on a ledge or beside a cliff facing the north. There is no reason to hurry. The only urgency is to get moving again in midafternoon early enough to stoke up for the night. Feeding continues until dusk begins to draw down, and then the animals bed down for the night. Water is seldom if ever the worry it is for big-game animals in many other habitat situations, nor are regular trips to a drinking spot necessary. Snow or snowmelt puddles or rivulets are always present.

## Movements

Day-to-day movements concerned with foraging are not ordinarily extensive. They are directly related to the amount of food available. Goats have been observed spending several weeks in the same area, and seldom more than a half-mile from some apparently central resting point. If food becomes scarce, they wander more. In fact, most of the movement of goats during any given day, and over an entire season, is dictated by the food supply.

In a few instances of transplants of a group of goats, the animals have wandered far. One such group was "lost" by observers who flew over the region looking for them. They were thought to have succumbed, but two years later they were discovered inadvertently when a plane survey of other game was made. The animals, with newborn additions to the group, had apparently followed high spines for many miles and were in another area entirely. Why this migration or wandering took place, no one knows.

In fact, there is still much to be learned about the habitats and life history of the

*Both sexes have beards on the lower jaws and long horse faces. The thin, black stiletto horns are used by males in dominance fights; either sex might use them on rare occasions against predators. But any goat's main defense is its climbing ability.*

mountain goat. Its home among the high crags makes constant observation over long periods difficult. Seasonally the pursuit of forage generally takes the animals lower down from their summering area. Deep snows are the cause. As noted earlier, they may move up periodically, to find places to forage where wind has cleaned the snow away. But the general movement is usually downward.

It is slow, simply a matter of seeking less snow on south-facing slopes where food is more easily available. This is definitely not a migration to a specific and ancestral winter range. If a winter happens to be mild by high-altitude standards, the goats stay put. In a few places to which they've been transplanted there is no need to move much lower at any time, and so they are found year-round on the same areas of a peak.

As a rule, the beginnings of spring find them at their lowest altitude. Here late-winter food is usually most plentiful, and here the first greening of spring begins, and gradually moves up the slopes. The goats move with it. However, this entire trek may cover only 1000 to 2000 feet in altitude, with side wandering along the ridges. Most observers believe that 10 or 15 miles in a year covers the range of the average animal. Some may be born on a high mountain and never travel more than 5 or 10 miles in any direction in their entire life span.

On occasion the need for salt urges a group into a fairly long, periodic trip. Game biologists have found what are apparently ancestral mineral licks that have been used possibly for centuries. In Montana, for example, such a spot was discovered by a big-game live-trapper high above the Middle Fork of the Flathead River. He built a pole trap here, with a trip wire to close the gate, and baited with a combination of salty minerals concocted by first testing what the earth contained here that had so long attracted the goats.

Soon animals were coming regularly, and were trapped. Some were marked with colored streamers so they might be observed later perhaps, to find out how far away they lived. The trapper came to believe that several groups of goats from various places in these mountains had long utilized this spot. Some of the marked goats were later seen, apparently "at home," almost 20 miles away. Whenever they felt the need for salts, they made the trip to the lick.

To illustrate the difficulty of trapping and transplanting goats, and why so little of it is done, the ones caught here were subdued cowboy-style, and a U-shaped length of garden hose forced over the horns. The horns are vicious weapons. A big goat could easily kill a man who did not use utmost care. In a pen a billy has driven a horn through a 1-inch pine board. Sturdy mountain horses brought to the site, which was 28 miles from the nearest road, were fitted with specially built twin panniers, one hung on either side. A securely roped goat was placed in each.

Down the mountain they went by horse a short distance to the edge of the brawling Flathead River. Here they were transferred to a big rubber raft, and floated downstream in a hair-raising dangerous run to a ranger station back in the mountains. At that point a small plane flew them out one at a time to a lower-country pen where they were held temporarily and then trucked to the new release site.

In bodily movements the goat is a most deliberate and seemingly careful mountain animal. It is not much of a runner even when pressed. When disturbed it begins a lumbering trot and then breaks into an awkward gallop. It covers ground at fair speed, but relies chiefly for safety upon escape into upended cliffs where pursuit by an enemy is hardly possible.

In ordinary travel, each step in broken areas is selected with care. Moving at a stiff walk, the animal makes certain the trail ahead is maneuverable. To a goat, however, that can mean a narrow ledge with sheer rock wall above and below. It traverses such places slowly, and also will walk like a tightrope artist along a shard of broken cliff to pass across a spot where no other passage is possible. Often in a tight spot the goat will turn to face a cliff, reach up with forelegs and hook its feet over a ledge above, then pull itself up. Or it may back very carefully and slowly off a ledge it has started to walk along where it discovers a dead end. Most interesting of its maneuverings is the act of turning around on this kind of ledge. It rears up, and standing on hind legs only, faces and presses against the cliff and slowly turns, then drops again to all fours.

Although its running gallop is slow, when the mood strikes it the goat can make surprising leaps for its bulk, from crag to crag or across rough places, covering 10 feet or more. Old lone billies love to pick out bedding places that would make a human observer shudder—sometimes pinnacles barely large enough to lie down on, occasionally so small part of the animal hangs off or overlaps.

Nonetheless, goats do make mistakes. It is believed that falls are fairly common, and rock slides and avalanches kill a good many. In fact, probably these natural disasters are the main control on the goat population. As noted, a few fall to predators. Predation may be a factor, it is thought, when the animals are forced down into lower areas where forest begins, and where such large predators as grizzlies or wolves may lurk. However, a mature mountain goat is a determined fighter when the need arises. If the stiletto horns can be slammed into an enemy, the goat is the victor. There are records of bears killed by large billies and of wolves put to rout.

When moving downslope in steep places, a goat commonly passes over near-perpendicular slants. It does so by jumping downward from one tilted rock face to the next, the grip of its splaying toes and cushioned hooves easily holding its weight as long as some momentum continues. Because of their experience in severely broken rocky country and their proclivity for passage in precarious places, when they do get down into timber, and into areas where blowdowns lie in jackstraw piles, they rather comically often walk the logs some yards above ground rather than trying to push through from below.

Aside from the daily feeding travel, and retirement to caves or other protected places during heavy rain and to cool places to rest, the mountain goat is basically a homebody. It may wander around looking for a dry spot of fine earth, paw out a small depression, and wallow in the dust, or stiffly plod along to a puddle for a drink, or make a trek to the salt lick. But as long as forage to fill its belly is present, the goat has little interest in discovering what is on the far side of the mountain.

## Breeding

Of course a billy recognizes a nanny, and vice versa, but hunters have long known that it is all but impossible to distinguish between the sexes at rifle range or even closer. It is so difficult and chancy, in fact, that it is not possible to set seasons on males only. A group of goats, with some individuals smaller than others, is probably composed of females and kids. A large, lone animal with an especially pronounced hump is probably a mature billy. The horns of the female are usually shorter and more slender, but unless comparisons are possible, the modest size of even the largest horns make correct sex judgment uncertain.

The problem of distinguishing between sexes is one reason that the hunting season for goats is almost always set in late summer—August—and very early fall. Throughout summer most of the adult males have stayed by themselves, usually as loners, although there can be exceptions. Thus a big lone goat is probably a billy. But when the rut comes on, about November and running on into December, they leave solitary ways and begin to seek companionship.

Evidence of pugnaciousness now shows in the male personality. A male may rake bushes with its horns, and brace any other male that crosses his path. At the rear base of each horn there is a gland that secretes a sticky, odorous substance. Both male and female have these glands, but those of the male are larger, and now as the rut begins they become distended. The goat rubs the glands against rocks, presumably to mark his presence.

Battles between males are not usually severe. Occasionally an especially crotchety billy stabs a sparring antagonist fatally. Most battles are simply simulated. Antagonists walk stiff-legged and haughtily around each other, spar a bit, and then break off. But the horns are used dagger-fashion meanwhile, and serious consequences may and do occur.

The billy makes no attempt to collect a harem. He simply moves in on a band of females and youngsters. If, as sometimes occurs, a mature male has stayed with a small band of females and young during summer, there is trouble instantly. But most of the groups—never very large, composed of three or four to seven or eight—are nannies and immature animals, from young of the past spring to males two to three years of age.

The billy may pursue a single female, or consort with two or three. He is possessive with these, but soon breeding is finished. Ardor cools. Bands may gather, now larger than at other times, to spend the winter. Or the same groups may continue to consort. Only the older billies seem inclined to revert to solitary ways again. They may hang on the fringes of a band, or go their own way again. Possibly the low incidence of severe battles among the males during the rut is a crafty part of nature's scheme. If the males battled as violently as, for example, bull elk often do, there is little question that with the goat's rapier armament at least half the males would be killed every breeding season.

## Birth and Development

Most of the females bred during any rut are not less than 2½ years old. Other may be as much as seven or eight, and possibly a few still older. Goats can be aged fairly accurately by growth rings on the horns, each ring designating one year. Although much is still to be learned about the life history of the mountain goat, a specimen with a dozen annual growth rings on the horns is considered old indeed. Goats held captive are not a good criterion. They never have done very well in captivity.

Approximately six months after the females are bred, the kids are born. A nanny living among a group draws apart and seeks a safe place for the birth. This is usually a secluded rugged spot hidden away from the others and from any possible enemy. Single births are the rule. Twins, however, are by no means uncommon. Some researchers believe the incidence of twins may be related to the quality of the habitat.

The little goats weigh 6 or 7 pounds and stand slightly more than a foot high at the shoulder. They are extremely well developed and precocious; they gain their feet often only minutes after birth. A kid may begin nursing its mother during the first hour of birth. Already it seems playful, jumping and walking around a few square yards of territory. By the time it is two to four hours old, it is able to follow its mother short distances.

*Kids are born about six months after the nannies are bred. Although the young are able to climb very well on treacherous goat trails soon after birth, they remain close to their mother until several months old. The kid in the background here never strayed more than a few feet from the nanny.*

The mother still clings to the rough hiding place as much as possible. Forage is likely to be good now, for the births occur in the May-June period, when new grass is beginning and there is a richness of juices flowing again in all available forage. The nanny is ever watchful, a determined and vicious antagonist if any predator attempts to take the kid. She stays away from the band for several days, but then emerges with her offspring and joins the family band she has left, or perhaps throws in with another.

No more than a week passes as a rule before the kid attempts to eat plants on which it sees its mother feed. Mountain goat kids are exceedingly playful, with each other and each one individually. Like their parents, they have no fear whatever of the high places. They will race along a sharp ridge or over the rocks without the slightest consideration for the awesome cliffs that fall away beside them. It is possible some young may fall to their deaths, but presumably such accidents are not common. Golden eagles are thought to prey to some extent upon the kids. Even an eagle attack is defended with spunk by the mother, or later on by any adults of a group.

If the band has had to winter some distance down the slopes, the slow movement toward the summit passes the short weeks of summer. Many groups, however, have not moved far if at all, and a fairly large gathering, of a dozen or more mixed nannies and kids and young billies, commonly spends an entire summer in a comparably small area. Observers in settlements far below have now and then spotted the same group high above day after day for weeks at a time.

By July and August most of the kids are fully weaned. They remain with the group, probably feeling the need for companionship and some sort of direction. The group organization is a loose one. Animals may scatter out to feed, seldom traveling in quite such meticulous single-file movement as is common among sheep. If some disturbance puts the band to flight, it's every goat on its own, selecting whatever path seems best to it. The kids may follow their own mothers, or by fall not be quite so closely tied.

Now fat and with winter hair growing, the young seem puzzled by the heightened activity of the rut. But once that time is passed the young animals and nannies winter together as usual. Some of the young males, even the long yearlings, may wander off by themselves. Others may feel stronger ties and stay with the band. Thus the simple life of this intriguing animal completes another cycle in this stark and rugged land that also is the fount of its simplicity.

## Senses

It may well be that this existence, reduced to basics, avoids the need for sharply tuned senses. Not that the mountain goat lacks a certain alertness and excellent development of the senses. But there is really no need here for the awesomely acute sense of smell with which the whitetail deer is gifted, or for the eaglelike eyesight of the pronghorn. Or perhaps a better view is that, though the goat has the keenness of fundamental senses, there is seldom urgent need to use them.

Researchers believe that the goat has excellent eyesight. Hunters agree. Yet what the goat sees, such as a moving hunter far off below it, doesn't seem very important. On occasion a hunter has reported getting close to a goat and having it move toward him out of curiosity. Unfamiliarity is not necessarily interpreted as danger. Goats fleeing from hunters often gallop only until they are well into the maze of broken crags. Nothing can follow them here, they seem to feel, and so there is no longer any hurry.

Hearing is also presumably well developed. But the high mountains are full of sharp noises—the clap of thunder, the falling of rock shards, the roar of a slide, the clatter of pebbles that suddenly roll from an ill-balanced resting place and set others free. Thus noises also are not invariably signs of danger. They are part of the natural phenomena of the surroundings.

Probably the sense of smell is also well developed. But here again there is not often much scent to catch that is disturbing. Further, the high peaks are alive with whimsical breezes that seem to blow from all directions at once. There is no reason why a mountain goat should be disturbed, for example, if it caught a whiff of a human intruder. A grizzly or a wolf would be different. It may have had experience, even ancestrally and the instinct passed down, with these. But a hundred generations of goats may never have smelled a man.

An interesting sidelight on the use of the senses concerns the daily habits especially of the big old billies. After feeding, one will move up to lie at the top of the peak or very near it. Daytime thermals rising below move any scent of danger up to it. Sounds, as mentioned, don't seem important, and those of falling rock or rolling stones are invariably below. The animal can see over a vast area below and around it. Thus there is very little to be afraid of or nervous about. Nothing can harm it from above—there isn't much of that! Unfortunately for the goat, many a crafty hunter realizes this and when stalking makes his climb to come upon the trophy from above.

## Sign

Hunters, photographers, and wildlife enthusiasts need not look very hard for signs when looking for goats. It is too easy to use binoculars or a spotting scope and locate the animals themselves from below by carefully scanning a mountain. However, one who has gone to the trouble to get up where the goats live should know what signs they leave.

Tracks are rather similar to those of mountain sheep. While sheep prefer a somewhat less rough terrain than goats, especially at their highest altitudes, both sheep and goat tracks may appear in the same region. The track of the goat is quite square, as is that of the sheep. However, the inner edges of the impression left by each toe are, to the practiced eye, straighter than those of sheep and the toes often splay outward farther. Length of an average adult track is roughly 3½ inches. Nonetheless, tracks of sheep and goat are easily confused.

Droppings are also easily confused with those of mountain sheep, and even with deer, although deer are seldom present in goat range. Those of the goats are generally a bit smaller. Like the others, they show different forms depending on diet. Summer droppings when the animals are eating grass and green weeds are in a rounded mass 4 to 5 inches long. By late summer when forage is harder, pellets may be seen but they adhere in clusters. By winter when browse and dry food make up the diet, separated pellets are formed.

It has been noted that goats hide out in caves in rainy or sometimes in extra-severe weather. Sometimes beds may be in shallow caves, too. Droppings and a definite bed with dusty perimeter in a cave or on a ledge indicate that goats may use or have used the area. But here again, sheep have similar habits. By and large, the most prominent sign of the goat is the white splotch of the animal itself showing against rocks of a high peak.

## Hunting

Every goat hunt begins by getting up into the mountains where the quarry is presumed to range, invariably on horseback, and then glassing for that most obvious sign, a lone goat or a group of them. To be sure, they are not easy to sight against snow. But the black horns do stand out to a careful, meticulous observer using a good glass. Further, seldom will the terrain be so entirely snow-covered as to give total camouflage to the animals.

The quality of the various hunting locations is roughly as follows. Alaska is best—that is, with the highest goat population, possibly as many as 15,000 animals. The goat does not range throughout Alaska, but is found chiefly in the mountains of the coast beginning in the lower Panhandle and reaching northward, and then to the area above Anchorage, and out in the Kenai Peninsula. Goats have been transplanted to Kodiak, and to some islands off the Panhandle. This is mentioned because all too often hunters have the idea goats are everywhere in Alaska.

British Columbia rates second. The Yukon and the Northwest Territories are both just fair, and the same is true for Alberta. Washington has a good population. Oregon a modest one, Idaho a fair one. Montana probably rates next to Washington. Colorado is estimated to have a goat population of several hundred. Other U.S. populations are token. And, of course, it is by no means easy to acquire a permit. Most are by application and drawing.

Almost all goat hunters are guided. A guide may be mandatory, depending on where one hunts. A guide is worthwhile whether mandatory or not. Selecting a trophy is tough for a tyro. If the horns appear to be as long as the distance from their bases to the end of the goat's nose, it is without question an excellent specimen.

Actually the only truly difficult part of goat hunting is getting to the shooting spot. It is seldom difficult to get within a range and still keep under cover, in the stand-on-end country of the animal's domain. The physical demands on the hunter are the tough part. When a goat is located, or a group, and a target is selected from far below, usually the technique is to watch until the billy lies down. Thus most hunts are launched early in the morning, and the animals are spotted while feeding and moving.

Once they have been located, the hunter can be reasonably certain they won't range far away. The hope always is that the hunter can spot where the chosen animal lies down along toward midmorning. It is virtually certain to be anchored right there for several hours following. The climb and the stalk then begin. Best technique is to circle wide to keep thermals from carrying scent upward, and to get above the bedding ground. Some expert archers have stalked goats in this fashion to within 15 or 20 yards with the animals wholly unaware.

On occasion after it has been decided that a certain animal is of trophy proportions, the stalk requires such a long time that midafternoon comes without the shot being made. And from the vantage point above it may well be that there are obstructions so that one cannot see the animal. Successful close-in stalks may still be made. But most hunters play it safe at this juncture. After the arduous work of the climb, they take no chances on flushing the quarry into terrain where they can't get to it. They wait until the animal arises and begins the afternoon feeding period. The chance is then good that it will move into view and offer a shot.

There are two main cautions in goat hunting. One is to be sure to carry a rifle with ample power to put the goat down to stay. These are tough creatures. Another is to watch carefully before accepting a shot, to make certain the goat won't tumble off a ledge and either become irretrievable or break its horns.

Whether one hunts the mountain goat or simply desires to observe it distantly or to photograph it, the knowledge that it appears secure and forever unendangered in its towering world in and sometimes above the clouds is a satisfaction. Further, even though this creature is seemingly one of the least alert and intelligent of our big-game animals, it has one quality no other can match. It is the most brilliantly surefooted of all horned and ant-lered North American animals. It is also uniquely and perfectly tailored to a world that undoubtedly will never be quite completely understood by man, because it is too big a challenge for all but a few to enter, and then never on fully intimate terms.

# Grizzly

*Ursus arctos horribilis*

# Alaskan Brown

*Ursus arctos middendorffi*

*A large male brown bear that may actually appear to be small in the water can suddenly stand almost 10 feet tall. The main reason for standing erect, probably, is to get a better view all around.*

**T**he grizzly bears are the most formidable and dangerous animals on the continent. In largest form—the Alaskan brown—the grizzly stands in size at the top of the world list of large carnivores. The only competition for that place it might possibly have is the polar bear. Without question the grizzly is the world's largest meat-eater living entirely on land.

Primitive Americans, early explorers and settlers, and modern hunters and wildlife enthusiasts all have been awed by, intrigued by, and curiously drawn to the challenge of these huge, unbelievably powerful creatures. No other North American mammal is so unpredictable, so irascible in temperament. None is so wholly without enemies of consequence.

The grizzly is the kind of wildlife personality that seems fashioned purposely by nature as a basis for endless legends, tall tales of the frontier and of the remote hunting camp. Indians from the earliest times revered the grizzly, chiefly because they feared it. The big bears figured prominently in various primitive rituals and beliefs. An Indian who managed to subdue a grizzly and take its hide was a legend in his time among his people, a hunter without peer, due the highest respect.

There is indeed a kind of super-macho romance connected with these great animals. Interestingly, no other North American animal quite so succinctly illustrates the rule of nature that the larger the animal, especially among those that eat meat, the fewer its numbers. And it illustrates another rule: The larger and more powerful and aggressive the carnivore, the more swiftly it is dominated and brought to severe decline my man.

Basically this is because of man's fear, both of the animal itself and of its depredations upon his stock, and because as civilization nibbles inexorably at wilderness there simply are very few places left where such a large and aggressive creature can make a living. Although over a substantial portion of their remaining range grizzly bears are still on the game-animal list, there is no question that their long future is bleak.

Over vast areas of their original range they are totally extirpated. In remote northern areas of western Canada and parts of Alaska they are still numerous and not presently endangered. Trophy hunting will not bring the grizzly to extinction, because at least in these modern times hunting can be meticulously controlled, or outlawed altogether, which well may occur before the next century. Usurpation of the remaining wilderness areas that are presently home to the grizzly is the real danger. And unfortunately, unlike horned and antlered game, the grizzly cannot very well be transplanted to new areas to launch fresh regional populations. It is not that adaptable, and besides, it has nowhere else to go.

When the first white men of colonial times had pushed westward into what are now the central states, they discovered the grizzly as far east as Ohio. It is easy to imagine the awe and fear of pioneers, already acquainted with the occasionally dangerous black bear, who suddenly came upon one of the great grizzlies of the interior. The Spanish had already found the grizzly well down the Sierra Madre mountains and their brushy foothills in north-central Mexico, and later white explorers were to encounter them there in numbers. They also inhabited Baja. The Mexican grizzlies were the smallest of the lot, but had a reputation for irritability and ferocity. This same grizzly ranged northward into portions of our present-day southwest.

Grizzlies were well entrenched all along the Pacific coast, throughout all of the Rockies, and over the Great Plains. Interestingly, although the big Alaska browns were not known during early American history, it is believed that the bear called at one time the California grizzly was possibly the largest and most powerful of all the grizzly tribe, at least south of Alaska. The plains grizzly, which was actually the same bear, was also large and powerful and dangerous. This is the grizzly that followed the buffalo herds and was known to kill an adult bull with one smack of a paw and a bite in the neck. Early westerners were forced to brace many of these giants, and a goodly number of men were mauled or killed in encounters.

It did not take long to decimate the grizzly population below the present Canadian border. The demise of the buffalo took most of the dependent plains grizzly with it. Later, stockmen and the so-called sodbuster farmers obviously could not tolerate such a large predator, and accounted for the stragglers. Grizzlies were gone from much of their original range within the contiguous states a hundred years ago. It is believed that within the past fifty years there has not been a drastic lowering of the population in that region. But everyone recognizes that very few are left, and those in only a few places.

In mid-1975 the U.S. Fish & Wildlife Service listed the grizzly within lower-48 borders as a threatened species. That classification pertains to an animal that may in the foreseeable future become endangered. A meager scattering remains in the Rockies in Colorado and north-

---

### THE GRIZZLY BEARS

**Color: Grizzly,** varied, from near-black to blond, usually brown with silvery-tipped guard hairs. **Alaskan brown,** varied shadings from very dark to blond, but without silver-tipped guard hairs, most specimens brown to tan.

**Measurements, adult males: Grizzly,** overall length 6 to 7 feet average, some slightly longer; height at shoulder to 3½ feet. **Alaskan brown,** much larger, 8 to 9 feet long, 4½ feet at shoulder.

**Weight, adult males: Grizzly,** from 450 to 800 pounds. **Alaskan brown,** 800 to 1200 pounds, occasional specimens to 1500-plus.

**Females:** Smaller, roughly at maximum about the minimum for adult males.

**General attributes:** Dished-in facial profile and hump over shoulders distinguish the grizzlies from black bear; very short tail; extremely long claws on forefeet (longer than hind claws), longer and less curved on Alaskan brown than on grizzly; awesomely powerful build; prominent, heavy, curved canine teeth; aggressive; always potentially dangerous.

---

## *Range of the Grizzly and Brown Bears*

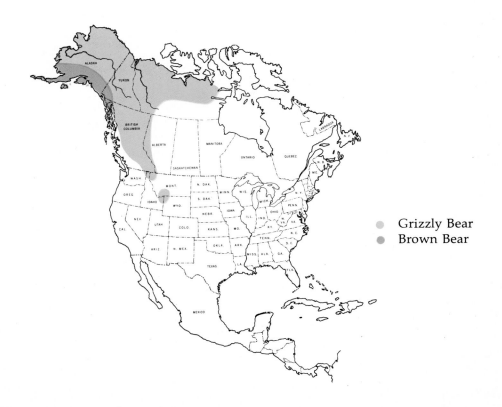

● Grizzly Bear
● Brown Bear

*Lower slopes of the Outer Range in Denali National Park, Alaska, are often shrouded with fog. They are home for many bears called Toklat grizzlies, which are blonde to cinnamon.*

ward, but most of the remaining animals south of Canada are in three regions: what may be termed the Yellowstone ecosystem, which takes in Yellowstone Park plus parts of Wyoming, Montana, and Idaho surrounding it; the Selway-Bitterroot region of Idaho and Montana; and northwestern Montana.

Token hunting south of Canada may still be allowed for a few years to come. Outside of controlled areas, like Yellowstone and Glacier National Parks, the only significant grizzly population from a hunting viewpoint is within the Bob Marshall Wilderness Area in Montana. Presently in that wilderness and surroundings the annual kill, whether by trophy hunting or by known poaching or for predation control, is limited to twenty-five bears. In 1975, to push protection of the grizzly south of Canada, the Boone & Crockett Club, keepers of official records of big-game animals, made a ruling that grizzly entries from south of the Canadian border would no longer be accepted. The Mexican grizzly, incidentally, is either extinct or on the brink of extinction. A few, perhaps two dozen, may be left.

North of the U.S. border the picture is bet-ter. British Columbia has a substantial grizzly population, and western Alberta has a fair number in the northwest, in the Peace River and Smoky River regions, and along the east-ern slope of the Rockies. The Yukon Territory is home to a stable grizzly population, of a strain not as large as some others but known for its tough cantankerousness. In the North-west Territories the animals are unthreatened. Alaska has probably the highest population,

*This fine grizzly in heavy fall coat is browsing on low-growing blueberries and crowberries on an Alaskan mountainside. The bruin eats not only the berries, but almost the whole small bush. This diet furnishes the animal enough fat to survive the upcoming winter while hibernating in an underground den.*

estimated at 10,000 to 12,000 animals, and un-threatened.

Scientific classification of grizzlies in Alaska has been confused for many years. At one time the confusion extended south of Canada. Naturalists and taxonomists split up the grizzlies into dozens of subspecies, chiefly based on geographical incidence. In early-day observations of grizzlies, color differences were to some degree puzzling to certain naturalists. One bear might be nearly black, but never with the sheen that the coat of a black bear often has. Another might have guard hairs conspicuously tipped with silver, or in younger specimens even yellowish, with the main coat dark brown. Still others ranged through various shades of brown to nearly blond. Eventually, however, it was agreed that the grizzlies south of the Canadian border were really all the same bear.

When interest in the huge brown bears of the southern Alaska coast became intense some years ago, and knowledge of them grew, they, too, presented a puzzle for scientists. Color among these bears also is highly varied, from very dark through varying shades of dark brown to tan to quite blond. But their enormous size seemed to set them apart. Scientists insisted that these bears of the Alaskan coast, and their sometimes even larger relatives on Kodiak Island, commonly called the Kodiak bear or Kodiac grizzly, were of at least one and possibly more than one species. Along the fringe of range between the outsize coastal bears and the inland dwellers there is, of course, intergrading. For some years it was anybody's guess which bear was which.

Presently it is acknowledged that all of the grizzlies are of a single species, *Ursus arctos.* However, most references, including the Boone & Crockett record book, which could hardly allow the enormous coastal browns to compete with smaller inland grizzlies, consider that there are two subspecies: the Alaskan brown, *U. a. middendorffi;* and the grizzly, *U.*

*a. horribilis.* The latter ranges inland throughout most of Alaska except the extreme northern coast and the southern coast. On the southern coast in a swath along the Alaskan Peninsula and on coastally southwestward and south to the border, and throughout some of the coastal islands, is the domain of the largest of North American bears, the Alaskan brown.

Oddly, the Alaskan browns are not commonly as irritable and aggressive as their smaller relatives, the grizzlies of the interior. Certainly both are potentially dangerous. But the smaller bears inland are often downright ugly in disposition. They have been known to stalk hunters, and to rush to attack with no special provocation. Fishermen after salmon and trout in streams along the southern Alaska coast sometimes carry whistles and blow them to alert brown bears along the streams to their presence. Usually the animals will move off when disturbed. It is important not to surprise one at close range, which will almost certainly spark a charge.

Only a comparatively few people, mostly hunters and wildlife photographers, have seen the big browns and other grizzlies in the wild. It is difficult to conceive of their astonishing size and power. A big grizzly habitually rears upright, standing tall on its hind legs, when alerted and uncertain of what has disturbed it. In that pose, it is 8 or 9 feet tall, and if it were to reach up to claw at a tree it might leave claw marks 12 feet from the ground. The hide from a large specimen, when spread flat on the ground, may measure as much as 9 feet from claw tips to claw tips of the front feet. Each claw is at least 4 inches long.

In the records, trophy grizzlies are judged by skull dimensions. Several skulls of Alaskan browns in the record book are almost 20 inches in length, measured without the lower jaw. Most are close to 18 inches. Width runs from 10-plus to very nearly 13 inches. Imagine adding flesh, hide, and fur to such a skull. It would hardly fit into a bushel basket. All the

*During the summer, bears shed their old hair for new, which by autumn is sleek and shiny.*
*This brown bear has apparently hastened the shedding process by rubbing up bald patches.*

record-book specimens, incidentally, are males.

Men have always been mesmerized by the power of these great carnivores. Notes of explorers and others tell of observing big grizzlies carrying off large kills such as bull elk and buffalo. Several hunters together cannot drag a downed bull elk, but a large grizzly handles one easily. Many years ago the early Spanish in America were intrigued with the idea of matching grizzlies with bulls. Near the earliest settlements in what is now California, it is said that bears for the purpose were roped from horseback by several riders—which must have been a dust-up to put any other horseback endeavor imaginable to shame—and dragged to enclosures. The result was, according to historical record, generally disappointing. Seldom was there much of a contest. A slap of the grizzly's paw often smashed the bull's skull on its first charge. A half-dozen bulls might be dispatched by the same grizzly as fast as they could be turned into the arena.

Much has been made in fiction and in some less than authentic nature writing about the roaring and bawling of the grizzlies, which would seem to imply that these are noisy animals. It is true that fighting adults do snarl, and that one making an attack may roar, probably a natural sound of anger and ferocity. Wounded grizzlies, and sometimes those in high anger, utter a frightening bawl. A disturbed grizzly may snort and snuffle as it tests the wind or eyes some possible interloper. It will growl menacingly, utter low woofs and coughs, and pop its great teeth together in anger or suspicion. These sounds may presage a charge. Cubs whine and whimper fairly often. But most of the time grizzlies are quiet animals.

They are also for the most part loners, except for female and cubs or yearlings together, and during breeding season. Each goes its wandering way. When several bears happen to come together, as in a stream to catch salmon or on a slope lush with wild berries, they appear to try not to get in each other's way, and even pretend not to look at each other. However, there is definitely a caste system. When an old, outsize male appears, the others move off, unless another male elects to dare the king to battle.

## Habitat

Except when it is necessary to get together because of a forage concentration or the mating time, there is no reason for the big bears to consort. Such large animals need roaming room to make a living, and of course this requires a large expanse of terrain that is truly wilderness. Thus the habitat of the big bears is almost entirely beyond the day-to-day touch or influence of man. Obviously, the grizzly that happens to move into the fringes of man's domain, as on a mountain cattle ranch of the west, is certain to get into trouble killing stock, which triggers its own destruction. But most grizzlies have a strong affinity for privacy, and with such large animals this means living in the wildest and most remote places.

On its original vast range, the grizzly's habitat varied widely, from the hardwood forests and stream bottoms of the central states, to the wide-open plains where small thickets of brush along streams might make bedding locations, to the forested mountains of the Rockies. But as this range was whittled down the only bears left were those in the mountain regions. The grizzly (*horribilis*) is of course still today generally an animal of the high country.

One who has visited off-road areas of Yellowstone and Glacier National Parks can get a good general idea of what grizzly habitat is like within the small range still existing south of Canada. In portions of western Alberta and throughout much of British Columbia the habitat is not especially different, except that as one moves northward there are many more open slopes and areas of only scattered timber, or even of high, rolling mountain country where brush such as willow is abundant but trees stand only in scattered groupings.

In the Yukon and Northwest Territories grizzly country is like this, and so is some of it in Alaska; it is a mixture of timber and mountain meadows, of river courses dense

*Grizzlies can be found along glacial river beds, on brushy slopes and tundra, on rock slides, and on open foothills as shown here.*

with low brush fringed by larger growth, and in the far north range massive brushy slopes rising to towering jumbles of rock. The mountain sheep, the caribou, and the moose of the northwestern portion of the continent share their habitat with the grizzly.

It must be kept in mind that the inland grizzly does not simply select a place, a kind of habitat in which it likes to live. It can exist in today's world only in remote wilderness areas, but even more important possibly, these areas must be able to support the animal. They must offer a forage potential fitted to the needs of the bears. Thus, unlike the mountain goat, for example, which seems unable to live anywhere but among the highest crags, and the antelope, which is specifically tailored to wide-open plains, the big bear—as it originally demonstrated long ago—is capable of adapting to any of a variety of altitudes and widely varying habitats, as long as there is enough food available to keep its belly full.

The Alaskan brown bear is a case in point. The well-known writer Jack O'Connor once spoke of the Alaskan brown as simply "a fish-fed grizzly." Although all grizzlies are omnivorous, a diet mainstay of the coastal browns is fish, salmon and trout that run by millions into streams tributary to the ocean. This habitat is somewhat different from that farther inland, and certainly not at all like the Rockies domain of the few remaining lower-48 bears.

Here stream courses are dense with low growth. There are endless open slopes where deep grass and spongy moss grow. These lead, of course, to the uplands, where timber begins to take over. Yet here too there are awesomely dense places of tall brush and scrub trees. On the grassy slopes in some places a bear can be spotted at long distance. But it is not the terrain, the habitat, that ties the bears here. This race was able to evolve because of the stable food source. Because enemies are unknown—except for the occasional hunter—the open places hold no fear for the animals.

In other words, the grizzlies are hardly products of a specialized habitat. Very similar bears—authentic grizzlies of the same scientific name—are inhabitants of various mountain areas of Europe and Asia, and the forests and plains of Siberia. The only restriction on colonization of the grizzlies of the world, apparently, has been climate. They obviously prefer cool climes, although some, such as the Mexican grizzly of the Sierra Madre, tolerated warmer weather than most, and one variety native to the Atlas Mountains once long ago had a counterpart in Africa. The point is, specific habitat is by no means as important as sustenance. Theoretically at least, a grizzly could make do almost anywhere, if it had at hand the wherewithal to stoke its mighty furnace.

## Feeding

Although the grizzly is usually spoken of as a carnivore, because it will eat all the meat it can get, the fact is that if it were not omnivorous the animal could not have survived on earth for all the centuries that it has been here. There is simply not enough meat available, at least meat the bear is capable of catching or finding, to sustain so large an animal around the seasons.

Thus grizzlies make the most of what may be termed seasonal crops, such as berries, fresh green grass, runs of fish, eggs of nesting waterfowl, and large animals such as caribou when killable during a migration. It is obvious that the bulk of the forage will be made up of what is most abundant in a particular geographical area, or at a specific time of year. In Yellowstone Park, for instance, the few grizzlies present for some years made a regular restaurant out of garbage dumps. This forage was plentiful for some months during the heavy tourist season. So-called renegades have been notorious since early settlement as stock killers. They learned to kill sheep, for example,

*The grizzly uses its formidable claws to tear open dead logs, feeding on rodents and grubs it finds in the rotting wood.*

or followed cattle herds, taking whatever they needed. It is no trick at all for a grizzly to move in on a range critter, catch it, and kill it.

Occasionally a grizzly may be out at any time in winter, traveling far to seek forage. These are exceptions. Most den up, and so the feeding year of the grizzly may be said to start in early spring—about April or early May—when it emerges, ravenous, and begins the search for sustenance. Now new grass is sprouting, there are roots and tubers to be unearthed, and varied kinds of tree bark are abundant and edible. The bears stuff themselves with vegetation. They graze much like cattle.

However, they are fundamentally meat eaters and the vegetation is simply a make-do. All the time they are grazing, they are ever alert for the scent of meat. In spring this is likely to be carrion, dead big-game animals that had too difficult a winter. The bears have no compunction about how ripe the dead moose or caribou. Anything available is used. This is also the time of year when numerous large animals

are weak, partially starved from a severe winter, a period when the old or ill make easy targets.

Of course any healthy animal—deer, mountain sheep, caribou, elk, or moose—that happens to let down its guard in bear territory may be rushed and slapped down before it can flee. In a race the bear, though fast for short distances, is no match for a healthy horned or antlered animal, but if it gets within close range its rush may be deadly. Spring is also the period during which the young of all big-game animals on the range of the bears are born. Grizzlies make a serious business of attempting to hunt down lambs, elk and moose calves, and deer fawns. No one knows just how effective these hunts are in terms of percentage of young devoured. Much depends on terrain and abundance of the prey. Undoubtedly, where the bears are plentiful, predation on young of big-game animals is substantial. The bears may not be able to scent young calves and fawns, which exude little odor, from any great distance, but they sense that this is the time, and when parents are in the vicinity they crisscross the region intent on tender young prey.

Meanwhile grazing also continues. The coastal brown bears may forage far and wide to begin. They may find carrion on the beaches—dead fish and seals perhaps—or they may be up in the highlands seeking weak or dead big-game animals. All this adds to the diet of vegetation. Then as the spring turns to early summer the grass shoots up higher and most of the bears will be in the valley or the low coastal areas continuing to graze like cattle.

The inland grizzlies of Canada and Alaska are fond also of small mammals such as ground squirrels and marmots. These are present in large numbers on the slopes. They are out for only a few short months during summer, before going back below ground again. The bears actively seek them, digging them out, even after they have hibernated. The long claws of the grizzly are perfectly designed for digging, and for snatching hapless small mammals from under rocks or among talus. An entire slope may be pocked where a bear has diligently hunted ground squirrels. It is meat that the big bears prefer.

When a large game animal is killed or a dead one discovered, the bear eats its fill and then caches the remainder, commonly scraping earth over it. Brush or sticks are also sometimes used. The bear is not choosy. Dirt serves as well. Gorged, the animal then retires to a nearby thicket if one is present, and lies down to rest, digest its meal, and watch the cache. It may even flop down atop the mound of the cache, and woe to any interloper that comes that way. Black bears often share range with grizzlies, but they have a healthy respect for their outsize relatives, and will flee at sight or scent of one, whether near a food cache or just in passing.

In certain habitats in fall grizzlies are able to feed heavily on acorns, piñon nuts, and other wild nuts. Bark of alder and willow also is

eaten in quantity as the animals begin to lay on all the fat they can. In other words, the big bears are opportunists, grabbing any good chance that comes their way. All of them eagerly consume fish, but it is the Alaska browns along the coast that are most definitely tied to the runs of salmon and trout.

In a few places there are spring runs of rainbows inland, and of suckers and numerous other stream fish. But by and large, fish are more incidental to the diet of the inland grizzlies, whereas the coastal browns' livelihood to a large extent depends seasonally on the fish runs. It is even conceivable that their great size evolved over many generations partly because of the rich fish diet they have always had at hand for several months each year. They not only attend the fish runs, but also hunt with great concentration all over the tidal flats, when tides recede, to pick up the bounty of marine life left stranded in tidal pools or on the mud flats.

By about the middle of July the salmon come into the coastal streams by tens of thousands, and there are also runs of Dolly Varden and other trout as well, some from out of salt water and some farther inland up the streams. The huge bears know the timing and begin gathering along the rivers and creeks as the first fish show. They continue to congregate as the runs come to peak. Air surveys have shown at times as many as fifty to a hundred big bears in a single watershed network near the coast.

Many written accounts have spoken of the big browns as rather nonaggressive animals. This would be wryly amusing to many a salmon-cannery worker or official tally taker of fish in coastal streams when runs begin. In some cases these officials have the job of counting salmon in certain runs and setting the time when the commercial fishermen may go to work on them. None of these people, and few others who have lived in the range of the big browns for years, consider them the least bit

*Brown bears often gather to gorge on migrating salmon. Although food is plentiful for all, there is often conflict among these normally unsociable animals.*

docile. Annually almost without fail there are maulings or deaths from encounters with the enormous bears.

Most of these occur on the fishing streams during the heavy feeding on salmon. Some are due to irritation of a sow with cubs at sight of an intruder. Some others are simply due to the short temper of an individual bear. They do not even truly associate with each other while fishing. Several may be working a stretch of stream, but each—except females with young together—stays to itself and seems to try to avoid even noticing another, as if each is attempting to stay out of trouble. Invariably there is an outsize male somewhere in the vicinity. When this king bear appears, the others scatter, full or not, unless some irascible upstart foolishly elects to make a challenging stand.

The salmon are caught by bears standing in the water, on a shallow riffle which puts the fish at a disadvantage, or below a falls which stops the run or concentrates the salmon before they leap it. The agility of the bears as they splash and whirl and feint at fish is astonishing, considering their bulk. Individual bears may do their fishing with slightly differing techniques, but the consensus among observers is that they do not swipe a paw into the water and flip a fish out onto the bank, as has been shown in some artist conceptions. Rather, the salmon is pinned between the forepaws and then seized in the mouth, or it is simply seized in the awesome jaw as it struggles up a riffle.

Smaller bears working a stream already fished by others ahead may pick up fresh scraps and leavings in their eagerness to feed, and then finally begin fishing. A big bear will put away six to a dozen husky salmon before

slowing down. Some especially gluttonous individuals, after sating first hunger, merely chop out a big hunk in the middle of a fish, drop the leftover ends, and go to a fresh catch. Others seem to have an uncanny eye for the roe-filled females, discard or pass by males, and seize and rip the bulging egg sacks from the female's belly.

Some of the bears eat while standing in the water, others catch a salmon and carry it ashore. The caste system among a group based apparently on size and age, dictates that when a stream location is being fished by the old male tyrant or king of the clan, others beneath him in stature must wait, or move to a new location. Observations by numerous scientists, photographers, and hunters indicate that the salmon, which die after spawning, are not eaten as carrion. They may litter a stream bank and fill the air with their stench, but the bears then spurn them, and move off seeking other forage. Some runs continue well into the fall.

## Movements

Most of the movements and travel of grizzlies are concerned with the daily task of getting enough to eat. The individual range of each animal is directly related to the food supply. Bears after salmon, for example, may stay within a small area for some weeks, making their daily trips to fish. When the slopes are a welter of wild berries, a grizzly may have no need or urge to wander, foraging day after day

*The fishing techniques of brown bears vary greatly. This one caught a salmon in its mouth and carried the fish out onto a gravel beach to eat it, eggs and all. Some bruins are much better at fishing than others.*

over possibly no more than a mile of aimless travel as it gorges on the fruit. But when such seasonal supplies are gone, the animal must broaden its search.

In a habitat where a daily living is not too difficult, the average grizzly probably does not roam over more than a plot of 8 or 10 square miles, possibly less. In shortage seasons, it may wander much more widely, within a circle having a diameter of 25 miles. Further, adult males may move about rather aimlessly, climbing a high ridge, retreating to a valley to potter along a stream course, almost without noticeable direction. When first out of the den

*Claws of adult grizzlies and browns are too long and insufficiently curved to be suitable for tree climbing. But they serve well as tools for digging out burrowing animals, tearing apart stumps, scooping fish, and ripping into other large mammals.*

in early spring, and again late in fall prior to denning, some bears may have to roam widely to find adequate food. However, as a rule before denning up the bear is larded with fat to tide it over winter, and thus may not be inclined to waste energy on aimless exercise.

There is some question whether the grizzly should be classed as diurnal or nocturnal. It is a little of each. Given ample easily acquired forage, most movement is by day. By dusk the bear finds a bed in willows or a conifer thicket, and sleeps until dawn. In lean times, or when grizzlies live in proximity to much human disturbance or intrusion, they feed and move around as much by night as by day, and possibly more by night.

All grizzlies, inland and coastal, are fond of following trails. Old roads, if present, and game trails may catch their attention. Soon these become bear trails. Some lead arrow-straight for hundreds of yards. If more than one bear is in the area, as is common in the coastal Alaska range, animal after animal may follow the same trail, as range cattle do, until it is worn deeply into a slope. However, the grizzly is in no way migratory. It does not change from summer to fall and winter range, as, for example, elk or mule deer may. It may switch its range simply because of a seasonal shortage of food, but the new one is never ancestral nor necessarily the same each time a shortage occurs. Again, these animals are simply opportunists.

The grizzly is so bulky that it appears, as it shuffles along flat-footed, to be a kind of immense bumbler. Not so! Angered or frightened, or in a charge after living food, or man, a grizzly can attain a speed as fast as a moose or even a deer—perhaps up to 35 miles per hour—though it cannot sustain this pace for long. It moves in a plunging gallop that appears astonishingly easy and even, considering its bulk, almost graceful. When an old male sets out to travel, perhaps with destination known, it may walk with plodding determi-

nation, unhurriedly. Or it may break into a trot, or an easy lope. Grizzlies have been observed running at that gait for several miles, seeming not to tire at all. Long, easy runs are not the rule, however. Unless put to flight, there is seldom that much reason to hurry.

Although the grizzly is a fine swimmer, only very rarely are there adults capable of climbing trees. This is because of their immense bulk, and their claws. Though long, the claws are not curved enough to allow a secure hold. Black bears, of course, easily go up trees; their much shorter claws are deeply curved. Grizzly cubs are known to climb occasionally. Ordinarily the claws of the inland grizzly are more strongly curved than those of the Alaskan brown. As grizzlies grow to adulthood, the claw curve lessens. Their great bulk would require a solid grip indeed to allow them to haul themselves up a tree trunk.

## Denning

One of the most purposeful movements of the grizzly is its early-winter search for a denning place. Bears are not true hibernators. Their metabolism continues at a rate only moderately less than before denning. They den up and sleep, but are not in a suspended state and can be aroused rather easily. A den may be prepared early in fall. If not, then by mid-September or on into October each animal seeks a proper place and makes its den. Most den sites are selected on a north slope. It is believed this choice avoids having water from melting snow run into the quarters, and also ensures that winter snow will plug the entrance and make the den snug and warm.

A cave may serve as a den, but usually the bear digs into the side of a slope, under an overhang, among blowdowns or under large tree roots. Some dens contain bedding material of branches and leaves. By the time the den is ready, or at least by the time the first hard storms are due to arrive, the bear becomes drowsy and is in a mood to curl up in its cave and go to sleep. Some researchers believe denning is triggered by the first severe snowstorm. The sleepy bears may be out and moving sluggishly around, their digestive tracts now empty. When the hard storm strikes, all go into dens at once.

Possibly so. Whether that is merely a regional occurrence that has been observed or a general pattern for all grizzlies, no one knows. It has even been suggested that grizzlies wait for the big snowstorm so that tracks to the den site are obliterated. Conceivably there may be some such instinct, each animal wanting to be sure it is well hidden away. It is difficult to believe, however, that fear is the impetus. They have nothing to fear from other animals. Possibly the bears realize a heavy snow will close them in. Or maybe it is simply that a light storm doesn't have any effect on a big bear, but at denning time a hard storm helps it make up its mind.

At any rate, depending upon latitude the grizzlies, wallowing in fat, perhaps several hundred pounds of it, are denned up anywhere from late October to November or early December. Not every one stays denned. Individuals may be aroused and do some winter wandering. Here and there a bear may den up only intermittently, coming out during warm spells. When April arrives, spring is on its way and the long sleep is about over. Late in the month, or in early May—again, latitude makes some difference in timing—the big bears awake and move out of their dens. They are empty, and the prodigious fat of fall is all-burned up, so the first urge now is to find food. By the time the transition to an active life has been achieved again, and the fat is partially replaced, the breeding season is at hand.

## Breeding

This is the only time of the year when adult bears consort. The females may be ready to be bred any time from early June on through the middle of July. Probably latitude influences the time to some extent. Not all come into heat simultaneously, of course. Further, females must be fully adult—that is, four or five years of age—before they breed, and each adult female breeds only in alternate years, some even every third year. Considering the fact that a grizzly, of either sex, has a life expectancy in the wild of probably not over twenty years, and usually less (though captive grizzlies have lived past thirty), the average female will breed only five to eight times during her life. This low potential production rate per female helps avoid overpopulation of so large an animal.

Although grizzlies are short-tempered during breeding season, it is known that very occasionally two large males may breed the same female and somehow avoid trouble with each other. Because the breeding season lasts several weeks, probably some males mate with more than one female. Much is still to be learned about the intimate habits of these large bears. However, individual males may continue a romance with a single female for several weeks.

During the mixing of the sexes, playful wrestling and nuzzling is now and then evident. The pairs forage together and show fondness and moderate affection. But as the season advances, ardor fades. The pairs drift apart, the older males once again becoming crotchety, irritable loners living as far from others of their kind as possible within a given domain.

## Birth and Development

The development of young grizzlies—and black bears, too—from the time the mother is bred until birth is one of the most interesting arrangements that has evolved in nature. No one is certain just why the complicated sequence developed, though there are some at least reasonable theories. Most of these bears

*A person who has studied grizzlies first hand can tell much from their face and facial expressions, such as whether they feel secure or agitated. Clicking teeth together is a bad sign. Such a bear is probably going to attack or at least bluff with a headlong rush.*

over eons of time have lived in severely cold climates, and the breeding season is in spring to midsummer rather than fall. The young, if born after what would seem a logical gestation period compared to many other large creatures, would therefore arrive in midwinter, in some cases even before the bears go into dens. Obviously, they could not survive.

This is especially true because baby grizzlies are unbelievally small and helpless at birth. They are blind—that is, with eyes snugly closed—and they are covered with short hair, brown as a rule for the Alaskan browns, and gray for inland grizzlies. The cubs number one to three, and in rare cases four; usually there are two. Coming from a mother weighing 400 to 800 pounds, each weighs at maximum roughly 1 ½ pounds, sometimes less. They are no more than 8 to 10 inches in length.

When the female is bred, the fertilized egg cells make only a start at development, which is arrested thereafter, and they then remain in a kind of state of dormancy, unattached to the uterine wall. That attachment, of course, is necessary before formation of the embryo can proceed. This unusual process, which occurs in a few other of the world's mammals, is termed "delayed implantation." Actually it is a process that assures at least a fair chance of survival of the young.

Sometimes in the late summer or early fall the uterine attachment is consummated and the embryos begin to grow. By this arrangement the female is certain to be denned up well before the helpless young are born. It is a kind of guarantee that the young will be born inside the snug den, and thus well protected. The gestation period runs somewhere between six and eight months. The female, having denned sometime between late October

*This sow brown bear is bringing twin cubs to a salmon river to feed. But she must be careful not to usurp the fishing territory of larger, stronger bears nearby. The cubs will learn their mother's fishing technique here.*

*It is always wise to give a female grizzly with cubs a very wide berth. Nothing quite matches the fury of a sow that feels her cubs are threatened.*

and mid-December, bears the cubs during the January-to-March period. By the time she leaves the den in April or May, the cubs, fed on her rich milk during the weeks in the den following birth, are ready to follow her, eyes open, more fur grown, and feisty as can be.

Some naturalists have theorized that the mother is so sound asleep in her den that she is not even aware of the coming of the cubs.

This is very doubtful. The birth of such tiny babies from such a huge animal is certainly not difficult, and just maybe part of nature's plan. But undoubtedly the mother cleans them and snuggles them against her in her drowsiness. Instinctively they find the milk source and really get excellent care—which is to say, just what they need, food and warmth.

The youngsters grow swiftly in the den, and

once out of it follow their mother and romp with each other, exuberantly playful, and continue swift growth. By midsummer they are ready to imitate their mother and try eating anything she does. She allows them to chew on ground squirrels she catches, and the coastal bears that fish for salmon carry the catch ashore to waiting and often frightened cubs, the big salmon possibly flopping vigorously. Soon the cubs overcome shyness, and once they've had a taste, they are well initiated.

By fall the young bears are sturdy, well filled out, and have grown a heavy coat of fur. They are constantly under close supervision of their mother, a ferocious, hair-triggered protectoress, from the moment they leave the den. Enemies are few. Oddly, the most serious danger is from other bears—males. An old male especially will kill cubs if it has the opportunity. In fact, trappers have sometimes used young bears as bait to catch old males. The mother, however, does not make it easy for the male. Even though she is much smaller, she instantly shows snarling hostility to any male that comes near, and will charge as quickly as she'd go after a human intruder.

The young of the year stay with the mother constantly on through the fall, now beginning to do some foraging on their own. They become fully weaned before denning time, when mother and cub or cubs go into the same den together. This also appears to be part of the grand plan. Compared to most other young, bears develop rather slowly and seem to need parental support for a longer period. Since the female breeds only every other year, or even skips two years, she can continue to mother her young into a second year.

The partly grown young bears continue to stick together and with their mother through the second summer. Some of them stay on until they are going on two years old, and a few even into the third year. On the average, the young den with their mother only during the first winter. That spring after emerging they begin to grow apart. This second spring the female is probably ready to breed again and start a new family.

## Senses

Experienced hunters are probably the keenest judges of the senses of animals. The successful hunter must thwart the animal's varied senses in order to come within range of it. Grizzly hunters have long known that the sense of smell is the first defense of the big bears. When they are searching for food, ever testing the breeze, they pick up the smell of carrion, for example, at extreme distances, and home straight to it. A whimsical breeze in mountain terrain that suddenly touches a stalking hunter on the back of the neck will instantly alert a bear at long rifle range and put it to flight.

Probably this sense developed with priority because it is the one most useful to the animal in making its daily living, and keeping in touch, regardless of the cover it may be in, with the world around it. Hearing is also sharp. But sounds do not disturb grizzlies as scents do, unless the sound is highly unusual. A bear foraging along a stream hears rushing water; one in the mountains is used to the sound of falling rock, or thunder claps. Even a gunshot, unless in the immediate vicinity, is not especially disturbing. The whistles carried by fishermen along coastal streams in Alaska where the big browns fish alert the animals and cause them to move off simply because this is an unusual, unknown sound. In some cases they may even learn to associate it with man's presence.

Eyesight is poor. The eyes are comparatively small. Undoubtedly eyesight has never developed highly for distance vision because it is used so incessantly for rather close work. Motion is instantly picked up by a watching bear, but it may not know, unless a scent drifts

to it, what the moving object is. A low-hunkered, stationary hunter who stays utterly immobile may be carefully studied and passed by. Unless a scent reaches the bear, it is not certain what the hunter is. At close range, of course, vision is excellent. But at close range the other senses also can zero in to bring multiple stimuli. All told, the sense of smell is the grizzly's keenest one, constantly whetted by use in its daily routines.

## Sign

It is obvious that an animal the size of a grizzly is certain to leave ample sign. In a few instances, where both black bears and grizzlies are on the same range, an inexperienced observer might confuse tracks and other signs of the two. Most grizzlies, however, are so much larger that there should be only minor confusion. And the differences between tracks in particular of the coastal Alaskan brown and the inland grizzly are distinguished not only by size difference but by the geographical location of each.

Bear tracks of adults leave, in good imprint material, the entire flatfooted impression of the hind foot. The forefoot track—the physical aspects of the forefoot are quite different—shows the toes and front pad, but does not always show the small, rounded imprint of the heel pad. It is the claws of the forefoot that are so long, in the grizzly. The length generally shows plainly, or at least where the tips make an imprint far out from the toes. This easily distinguishes them from the shorter, curved claw prints of the black bear.

Track measurements, of course, differ according to the size of the bear. Average adults should measure about as follows: inland grizzly, hind foot 9 to 11 inches to ends of toe prints, greatest width 5 to 6 inches; Alaskan brown, hind foot 15 to 17 inches by 10 or a bit more in width; black bear, hind foot 7 to 8 inches long by 3 ½ to 4 inches wide. An obscure fact little known to wildlife observers is that the so-called big toe and little toe on a bear are exactly reversed from the same on the human foot. In some materials the little toe barely leaves an imprint, and sometimes none at all.

A walking bear shows the hind-foot imprint a short distance in front of the forefoot print on the same side. Occasionally an individual bear when walking may overlap hind-foot and forefoot prints. At a gallop individuals may make slightly differing patterns. As a rule the hind feet are well out ahead of the forefeet; the two hind-foot tracks are in line, one behind the other, and the forefoot tracks are close together but slightly staggered off to one side. Looked at straight-on from behind, the galloping bear appears to run much like a dog, the rear end not quite lined up with the front end. However, gallop patterns of individual bears commonly differ.

As noted earlier, grizzlies like to follow trails. Most of these they make and continue to use, sometimes for years at a time, even following generations walking them. At other times a trail begins with an old moose trail or an old trail or road cut by man. Streamside trails are broad and well packed. Trails in grass, as on the coastal slopes or tundra, are much like old wagon trails. The feet on each side

*This large brown has been fishing but turns to stare for an instant at the photographer. Its scenting ability is much better than its eyesight, which probably is weak at long distances.*

*This Alberta grizzly's left paw is about 6½ inches wide and 12 inches long to the end of the claws. Note that the little toe and big toe are reversed from those on a man's foot.*

wear a rut and there is a ribbon of grass left in the center. Now and then a trail may be discovered where bears have meticulously stepped time after time precisely into the pad tracks they or other bears have left before. With much travel, these trails, especially in soft or moist earth, are rhythmically pocked along each rut, each track hole several inches deep.

So-called "bear trees" have been the stuff of frontier tales since the first grizzlies were known. Supposedly a big bear selects a tree in a prominent place near a trail, so other passing bears will see it, stands on hind legs, reaches up as high as it can, and rakes the bark. The next bear tries to outdo it, and so on. This, so the tale goes, leaves a kind of challenge, the biggest bear telling the others to beware. Of course this is just old-timer talk to impress the tyros. Bears rub their backs and bellies on big trees, often coming to the same one habitually. They rear up during the process, seize the trunk and rake it, and even bite at the bark, leaving tooth gashes as well as claw marks. After a tree has been rubbed and clawed, another bear, snuffing around it, may also use it, just as a dog or coyote sniffs at a tree where another has urinated, then does likewise. This is how the bear tree is established.

Trees with bark ripped away, chiefly conifers, sometimes with the trunk entirely girdled, are a different kind of sign. These indicate that a bear has been after the soft layer and its oozing juice below the outer bark. On occasion one may also find a grizzly bed. These may be simply lie-up spots in brush, or they may show that the bear has raked together branches or moss to make a soft place to take its ease. Dens may be spotted occasionally, but most are hidden away where a casual observer is not likely to find them. Plowed-up areas where a grizzly has been digging tubers in a low place or digging out a colony of ground squirrels on a slope are both obvious signs, the latter plain enough in the open to be spot-

ted through binoculars at a considerable distance. Food caches have already been mentioned. When one is located, it's a good place to be shy of unless you are hunting, and then it pays to be alert, for close examination may bring a charge from a watching owner.

Droppings also are easily spotted sign, especially on feeding grounds where an animal has spent some time. Droppings are large, and take several forms. Fruit, grass, and fish diets will make them formless and very soft, reminiscent of cattle manure when they're eating fresh green grass. A meat diet will show lots of hair in a much firmer scat. Such droppings are long and round and over 2 inches in diameter. All of the signs mentioned are ones wildlife observers, photographers, and hunters should be aware of. Fresh sign of several varieties means that one or more grizzlies are certain to be using the area.

## Hunting

Grizzly hunting is closely controlled everywhere nowadays. Because it is strictly trophy hunting, there is no sense whatever in taking a small specimen. Judging what is a top trophy and what is not quite is a job only for a competent, thoroughly experienced guide. Indeed, all grizzly hunts nowadays—this includes, of course, the Alaskan brown—require guide and outfitter services. Some hunts on the Alaskan coast are made by boat, most of them during the spring soon after the animals leave their dens. The craft, on which hunter, guide, and cook live, is the means of transport. Coastal areas are scanned by glass, and the hunter and guide are put ashore by wading or via a small boat to hunt on foot. Some other hunts in the interior are fly-in, and others via pack train. These may occur either in spring or in early fall.

The most important equipment for a grizzly hunt is a rifle of adequate power. Although many large bears have been downed by good marksmen with a rifle such as the .30/06 using a heavy bullet, nowadays the much more powerful large-caliber magnums are a far better tool. A hunter should give himself every chance to put the animal down for keeps with the first shot. A wounded grizzly hiding in willows or other cover presents a seriously dangerous situation to both hunter and guide.

Regardless of where the hunt is held, it is never easy. In mountains the climbing is arduous. Along the Alaskan coast, weather that is considered "good" may be by standards elsewhere abominable, always damp and chill. Walking in the moss and tall grass and at times through fantastically dense cover is exhausting. Twenty or thirty years ago it was easy enough to book a grizzly hunt into territory where little or no hunting had occurred, and bears were enormous and plentiful. One may now find areas where they are plentiful enough, but virgin grounds are all but gone, and because of cropping of the largest, oldest males, chances of record-size specimens are not as good as they once were.

As noted earlier in this chapter, no one is certain how long grizzly hunting will be allowed to continue. With miserly and astute management of the resource, it may still continue for several decades. There are attempts underway here and there to so regulate grizzly hunting that a sportsman would be allowed only one in a lifetime. A somewhat similar arrangement is in use for token populations of other large game animals in some states. If a hunter gets a moose permit, let's say, he cannot apply again, regardless of whether or not he makes a kill, for two, three, or five years, whatever the law states. Although the one-in-a-lifetime rule might be rather hard to enforce, it would certainly encourage true trophy hunting. No sensible hunter would settle for a so-so specimen, especially not at what it costs nowadays—several thousand dollars—to make a hunt.

There is nothing especially complicated about grizzly hunting techniques. If one is physically able to stand up to the hunt, it is a matter of riding horseback and glassing constantly, or walking and glassing. The open slopes, and the harder-to-scan low places with brush and tall vegetation, must be long and carefully studied with a glass to be sure of spotting the quarry. Guides always hope the animal will be at a substantial distance. That way the animal can be studied and sized up, and if it doesn't look big enough it is passed up. But if it is a shootable trophy, then a stalk can be planned to get within range, which means anywhere from 100 to 250 yards. Come upon at short range, a grizzly may charge, and no guide wants that to happen—not only from the standpoint of danger but because it may not be a desirable trophy yet might force a shot in self-protection. Also, at close range the bear may bound into cover and, alerted, get away or be much more difficult to track.

Of course, sign is always in mind. The places with the most sign get the most attention. However, because grizzlies are so large, and because the inland tribe particularly may be anywhere in a wilderness setting, a constant lookout and glassing are the chief methods of locating game. It is considered advantageous—and safer—always to try to get above the quarry during and at the end of the stalk if the terrain will allow it. Keeping the wind in favor is of course the main concern. With an animal finally decided upon and the stalk completed, the hunter should always take plenty of time to settle down, get his wind, and select an aiming point with utmost care before making the shot.

Grizzly hunting is a dramatic endeavor, always spiced with danger. The challenge of hunting so large an animal is undeniable.

Without question, however, the grizzlies still left on this continent have an uncertain future. Those of the far north undoubtedly will be here in at least modest numbers for many years in the future. But it can be reliably predicted that grizzly hunting will see more and more restriction, and ever more careful cropping of the existing population. Overhunting certainly can be harmful to the species nowadays. That is not likely to occur, because of enlightened management. The basic problem faced by these huge bears is one of wilderness living room. Man's so-called progress incessantly chips away at it. Whether there will be any of it left a few decades hence is difficult to predict.

# Black Bear

*Ursus americanus*

No other large North American wild animal has been more enmeshed in the skeins of the country's history and its legends, nor has any appealed more to American imaginations, than the black bear. The prowess of the colonial hunter was rated, in the final analysis, by his success with bears. Daniel Boone staked out a monument of sorts on the spot where he "kilt" his first "bar." Bear fat was used to grease boots, sometimes even hair and wagon wheels in a pinch, and even served as cooking oil. Hides made robes that were status symbols. Bear steaks and mulligan swirled their aromas throughout many a log cabin.

If boys were bad, a bear might get them. If they were good, they could sleep on the bear rug with the head and claws on. If the bears were exceedingly fat in fall, the winter would be long and severe. If they appeared from their dens early, it would be a mild and good growing year. The pigs, and old Bossy, must be watched on a crisp fall night to make sure bruin did not raid and carry them off. In summer beehives also bore watching, if there was to be honey for the home folks and not for the confounded bears.

*Probably no other large North American animal has been more enmeshed in the skeins of history and legends, nor appealed more to the American imagination, than the black bear. Often the prowess of the colonial hunter was based on his success with black bears.*

213

Nursery tales were filled with cute cubs, and with terrible boar bears hard to rhyme with that chased cool hunters who dispatched them with supreme bravery amid much black-powder smoke. Grandpa's bear stories held listeners enthralled by hearth and campfire. Cute, cuddly teddy bears have long been play-mates and security symbols for kids. And in modern times movies and TV shows have dis-graced beardom and their producers by giving bears human traits and intelligence and the gentlest of personalities.

There have been circus bears, dancing bears, leashed bears begging pennies for itin-erant fiddlers and accordion players, bears ex-hibited walking around and around smelly cages at roadside zoos. Then there was Smo-key, the orphan cub from New Mexico that grew up to become nationally renowned for urging citizens not to set fire to our forests. And far from least, there are the Yellowstone Park bears, which countless tons of printed flyers have warned millions of visitors for dec-ade after decade not to feed or fool with—and the term the same bears put into the language, "bear jam," which means a lineup of cars blocking the highway for anywhere up to sev-eral miles, formed by tizzying tourists feeding and fooling with them regardless, and occa-sionally getting cuffed and chewed on for their lack of sense.

It is little wonder that the black bear has always been such an important wildlife sym-bol. Although it is the smallest of the conti-nent's bears, it has always been the best col-onizer. In its adaptation to many varied habitat and climate situations, the black bear in some ways might be compared to both the whitetail deer and the cottontail. Almost everywhere that early settlers and explorers went, they en-countered this animal.

Presumably the black bear originated in Asia and entered North America long ago when, as geographers believe, there was a connection of land across the Bering Strait. Whether or not that is true, these bears col-onized all of Alaska except the far north and parts of the Peninsula, and all of Canada to where the forests disappear in the north, ex-cept minor plains expanses of the south-cen-tral prairie provinces. Within what are now the contiguous states, black bears managed to dwell happily and push forward practically everywhere except over the wide-open plains of the Great Basin, the treeless country of east-ern Oregon, and parts of the southwestern deserts. The original range continued far on down into Mexico, over both eastern and western flanks of the Sierra Madre and the pine and oak forests of some desert mountain ranges.

Obviously this vast range, to which the an-imals adapted over thousands of years, took in almost every conceivable combination of terrain and climate. The saw grass and palm hammocks of the Florida Everglades, the mountains of Maine, the hardwood and mixed forests of what are now the central and Great Lakes states, the Rockies and their broad foothills, along the Pacific coast except south-ern California—all these were home to the black bear. Climate apparently was no barrier, nor for the most part was altitude. Blacks are common in many places in the southeast in near-sea-level swamps, and also at 6000 and 7000 feet in all mountains. Hunters have bagged them and seen them much higher, to 9000 feet and above, well up toward the timber line, although probably they do not actually live at or above the timber line.

Because of its highly diverse diet and un-restricted tastes, the black bear can live—and has lived—practically anywhere. However, the one barrier to range expansion appears to have been where forests ran out. This is a for-est or woodland creature. Unlike the plains grizzly that followed the buffalo, the black bear, unless broad stream bottoms thick with

---

### THE BLACK BEAR

**Color:** Highly variable geographically; eastern Canada and U.S., glossy black, with or without white area of varying size on chest, muzzle and part of eye area paler to brown; western U.S., Canada, Alaska, cinnamon to brown to tan to occasional blond color phases common, plus black; Alaska glacier bear, a subspecies (U. a. emmonsii), smoky to iron gray to bluish to near-black, muzzle brownish; Kermode's bear *(U. a. kermodei), islands off west-central British Columbia, rare, white, nose grayish-sand, claws white, foot pads brick-reddish, eyes brown.*

**Measurements, adult males:** 4½ to 6½ feet overall length, rare specimens 7 to 8 or even 9 feet; height at shoulder, average 3 feet, some less, some slightly taller.

**Weight, adult males:** Average 250 to 400 pounds, exceptional specimens 500 to 600-plus.

**Females:** Smaller on the average by about one-fifth.

**General attributes:** Coat when at best has glossy sheen; very short tail; face straight, not dished as in grizzly; shoulders without hump as in grizzly; claws short compared to grizzly, and well curved; shy, not naturally aggressive, but occasionally dangerous.

---

## *Range of the Black Bear*

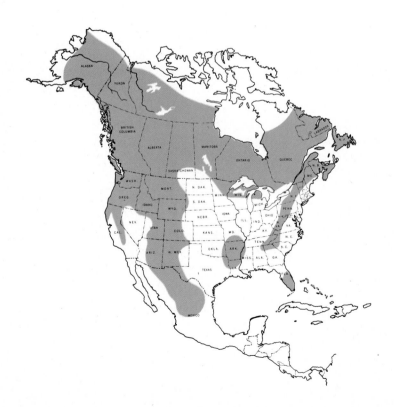

timber and brush were present for it to follow, never has been able to adapt to wholly open country.

Also unlike the grizzly, the black bear has been able to adapt rather well to man's presence. As an animal personality, this bear possibly has a keener intelligence and more cunning than its larger relatives. It is decidedly a shy creature, with an uncanny ability to keep its private life private. Its shrewdness in finding places to hide even on the very fringes of settlement is in many ways comparable to that of the whitetail deer. Today black bears live within short distances of very large cities, where habitat is suitable and food available.

In spite of its adaptability and its secretiveness, the black bear has suffered a serious population decline compared to earlier times. Certainly the species is not remotely endangered. But it has nonetheless disappeared from large areas of its original habitat. There are numerous reasons. In a few instances, particularly in colonial times and on to the mid-1800s, hunting pressure, chiefly hide hunting, depleted bear stocks locally. As human settlement progressed, bears in the vicinity of settlement were systematically destroyed because of their exasperating inroads upon livestock of all kinds, from sheep to pigs to cattle to poultry. Because blacks are extremely fond of sweets, bee raisers in many places—Florida even today is a classic example—always have had to take severe measures. A single raiding bear can put a bee man out of business in a night.

The most important influence on black bear populations was clearing of the forests, which drove the black bear from tens of thousands of square miles of its original range. It needs roaming and foraging room, and thus must have fairly large blocks of unbroken woodlands or wild lands in order to survive.

Today, black bear populations outside the contiguous states rate about as follows. In Alaska, very high; across the far-northern areas of Canada, substantial populations in the Yukon and Northwest Territories; in British Columbia, a good population; in the northern portions of Alberta and Saskatchewan, also a good population; in Manitoba, only fair to good; in Ontario, excellent; in Quebec, plentiful; in New Brunswick and Nova Scotia, good; in Newfoundland and Labrador, some.

In the eastern states, Maine has a high population, New Hampshire and Vermont are fair, New York is fair to good, Massachusetts and New Jersey have a paltry few left. Pennsylvania rates quite high—a surprise considering human population. Virginia has a fair number left in its western forests; North Carolina has quite the opposite pattern, with most bears in the tidewater region. South Carolina, West Virginia, and Tennessee have only a few animals left, and in Kentucky they are probably all gone.

The Great Lakes states—Michigan, Wisconsin, Minnesota—are still home to a large but slowly diminishing population. The Ozarks, once home to a good number, now hardly count. Missouri has a few, and Arkansas has made transplants but certainly isn't a bear state.

In the south, Florida still has possibly a thousand animals all told, Georgia a few, Mississippi possibly a few. Other southern states are questionable. Texas still carries the black bear on the game list but has none except a token population in the far-west Guadalupe Mountains.

It is in the Rockies and the northwest that the highest black bear populations remain south of Canada. Here Washington has the most of any lower-48 state—in some areas to nuisance proportions. They are anathema on timber-company lands, where they destroy hundreds of acres of young planted conifers by eating the inner bark. Oregon and Idaho have large populations, with Montana and Wyoming next in line. There are a surprising number still left in northern California, a few in Utah, and a fair to good scattering in Colorado, New Mexico, and Arizona.

*The color of black bears is highly variable, ranging from tan or cinnamon to the most common color, jet black. In the West, brown black bears like this one are common.*

Because of its vast range, the black bear was a perfect candidate for the "splitters" among the taxonomists to whack up its population into endless subspecies. They divided it into numerous races, based on minor physiological differences that were possibly the result of geographical and climatic influences rather than genetics. However, except for two distinctly different races, there is no reason to be concerned with these, and indeed it is probable that the majority of the continent's black bears are a single variety that needs no closer delineation.

In the days of early western exploration it was believed, with some logic, that the brown and tan bears so common particularly throughout the Rockies were of a different species entirely. The so-called "cinnamon bear" was the subject of much interest in those times. However, it was soon observed that a black sow might be trailed by a brown cub and a black cub, that shades of color were even more varied than had been thought, and that all of these were simply color phases of the same bear. Individuals can be light tan, dark tan, chestnut, cinnamon, and even pale blond.

*Living in a restricted range of a few islands off the British Columbia coast is a white- to cream-colored race of Kermode black bears. This rare creature is not an albino but breeds true to its unique pale color.*

These last darken with age. It is indeed interesting that very few brown black bears have ever been recorded from the eastern half of the country or from eastern Canada. An occasional one turns up, but it is rare. Blacks of the east, much more commonly than those from the west, often have a patch of white on the chest. This may be a snow-white area a foot across, or only a few white hairs.

The two striking exceptions of subspecies are northern. One, which lives on a few islands in a very restricted range along the coast of central British Columbia, is a strikingly beau-

tiful animal, often pure snowy white, with the muzzle sandy to gray-beige, the eyes brown, the soles of the feet reddish brown, and often the claws white. This bear, usually called the Kermode black bear, *Ursus americanus kermodei,* is not an albino, but a "sport" of sorts that apparently breeds true. It is a rare creature, fully protected.

The second notable subspecies is the blue, or glacier, bear, *Ursus americanus emmonsii,* also with an exceedingly restricted range. Its color is from pale smoke to bluish silvery gray to dark iron gray, and in some specimens a rich near-black gray. Its home is a section of coastal Alaska, in the St. Elias Mountains and along Yakatut and Glacier Bays. Named both for its color and for the fact that it lives amid numerous glaciers in awesomely rugged mountains, this handsome animal is elusive and rather rare. It may be less rare than supposed, since its forbidding habitat has kept hunters from collecting many and scientists from studying it in depth.

Whether or not the black bear should be considered potentially dangerous to man depends upon which writers you read and to whom you talk. Some writers and outdoorsmen scoff at the idea of danger from black bears. They claim, correctly, that blacks are shy and furtive, are wont to disappear like smoke at the slightest intrusion, would rather run from a man than stand and fight, and are quick to learn, unlike many other carnivores, from varied frights such as being chased by dogs or shot at by hunters. Therefore, they say, there is no reason whatever to fear them. But the facts hardly prove this conclusion. More people have been mauled, and killed, by black bears than by their large relatives. This may be because there are more blacks, and over a greater range. Nonetheless, it should impart a message.

To be sure, black bears are not very aggressive unless aroused. They are not as a species especially irritable or quick tempered. What they are is individually wholly unpredictable. Berry pickers have been hurt, and even killed by blacks. This may occur because the bear suddenly feels trapped or startled, or simply because it doesn't want any other creature sharing its forage. Sows with cubs are notoriously edgy and often vicious, which is understandable. Old bears, injured bears, even extremely hungry bears, are irritable and cantankerous. A bear startled at close range may fly into a protective rage. Any black bear deserves respect as potentially dangerous. One of the problems, in places like Yellowstone Park, for example, is that the average urban dweller whose knowledge of bear is limited to Smokey and cuddly teddy bears has no idea how strong and fast a bear can be. A yearling cub weighing under 100 pounds can knock a man flat with one easy swipe, or snap his neck with one bite of powerful jaws and long canines.

Along with unpredictability, black bears seem to have in them a kind of devil-may-care vandalism syndrome. They appear, exasperatingly, to take some perverse pleasure often in simply wrecking anything that comes handy. Combined with this is their curiosity. In campgrounds, aside from bedeviling garbage cans, blacks are prone to ripping up tents and smashing anything that lies around. In some instances food draws then and they bite into canned goods and smash ice chests. They commonly break into woodland cabins, and have been known to try this even while the owner is present. Like the trained dancing bears of the old roadshows, they are clowns of a sort, but they are not always funny.

Ordinarily the black is a quiet animal, uttering vocal sounds only when irritated, or making love, or mumbling to offspring. Large bears may bawl or roar, snuffle, make grunting sounds. When suspicious a black may rear up and snort and grunt. When it begins to cough, growl, and snap its teeth together, these sounds should be read as signs that it probably intends to run only one way—toward the intruder, with purpose. A wounded black on occasion utters a hair-raising moan. Little cubs may let out diminutive bawls when scared, and they whimper and whine to their mother. However, many a woodsman who has been in black bear territory for years has yet to hear one utter a single sound.

One reason may be that except for cubs and mother together, or male or female consorting during mating, black bears, like other bears, are loners. The adult males stay to themselves all year except for mating season.

On ranges near settlements, the big boars are inclined, probably because of long experience, to stay in the thickest, most remote cover, except for forays and raids outside. It may also be that bears are silent most of the time because they have no enemies among other wildlife, and thus little reason for bluff or rage. Large blacks may attack, and even eat, young blacks. But aside from man and his dogs, or the rare occasion when a big bear might be hurt while trying to down an antlered game animal or starve because of a mouth festered from porcupine quills, black bears are safe from hazards other than internal parasites. In territory where grizzlies occur on the same range, they brook no interference from blacks. The black bear that sees or smells a grizzly doesn't waste time growling or bluffing. It flees.

## Habitat

Perhaps the best way to define the fundamentals of black bear habitat is to say that this is an animal requiring cover, with at least portions of it dense. This includes a broad variety of range. Some bear hunters in Florida, Georgia, North Carolina, and Virginia might think of the bear as a swamp animal. The Everglades and Big Cypress, Okefenokee, in Florida, and Georgia, the tidewater swamps of North Carolina, and Dismal Swamp in Virginia have bear populations still today. Yet westerners think of the black as an inhabitant of mountain forests and meadows.

The black bear was originally so adaptable that it was able to live happily in both swamps and high mountains as well as in intermediate forest zones. The reason some of the southern swamps still are home to it, even though it may be scarce in upland portions of the same states, is that the near-impenetrable swamps are last stands, remote, with only moderate intrusion by man, and capable of furnishing enough food.

Black bears will venture from their woodlands into open areas, but not very far. It is common to observe a big black in the mountains of Wyoming or New Mexico moseying about in a high-country meadow, eating grass or remnants of a long-dead elk, or grubbing for small rodents. But these meadows are simply openings in the forest, each rimmed by mixed aspen and coniferous forest, Typical of optimum black bear range east of the Mississippi are the mixed forests of Maine, Pennsylvania, northern Michigan, and portions of Ontario, where pine, spruce, and other evergreens mingle with poplar, birch, maple, and varied oaks.

Brushy stream courses bordered by alder and willow tangles, or fringed with serviceberry, black haw, chokecherry, and similar fruit-bearing shrubs invariably are used heavily by bears within their forest domain. The dense thickets of spruce in portions of Canada and the jackpine and poplar plains and wild-hay meadows of the northern Great Lakes country also are home to the black bear. In the Rockies, the wild river courses and their hemming slopes, thinly or densely forested, serve perfectly as bailiwicks, and along coastal Alaska and British Columbia the streams with dense thickets plus the grassy slopes nearby sustain this bear in settings much like the home of the Alaskan brown.

Basically it is not so much the specific type of habitat that keeps the black bear contented as it is the amount of forage present. Given places to hide in peace, room to wander without too many frights or disturbances over a big block of country, the black bear can adapt to almost any surrounding—as long as there is enough food present to keep its belly full. Where man has settled the edges of large forest tracts, black bears don't mind at all living in proximity. They are drawn to these fringes because of the windfalls of garbage and other sustenance that result from man's presence. But a bear cannot abide a small woodlot, as a

*Black bears are quite at home in wetlands offering food. All bears are excellent swimmers and do not hesitate to hit the water.*

whitetail deer may. It needs room where it can make its presence either unknown or at least not obtrusive, and where, when harried, it can retire to tangled thickets and still eke out a living.

## Feeding

Given cover and room to roam without constant disturbance, food is the key to where the bears must be. Although classed as a carnivore, the black could not possibly have been so successful a species had it not adapted to a broad menu. Black bears will eat meat whenever they can get it. They are not the least bit choosy, either. A healthy live forage animal, a sickly old one, a dead one so ripe the carrion smell permeates the forest breezes for hundreds of yards around—the bear isn't finicky.

A nest of baby mice is slurped up grass and all. The nest of a wild duck on a marsh edge is gobbled, perhaps full of eggs partly incubated, or of just-hatched young; if a parent can be seized, so much the better. A newborn fawn, also its mother if a sneak attack is successful, an elk growing old and feeble, a pig from a forest-edge pen raided in darkness, the dog that barks at the raid, even a young bear of its own kind—all are accepted without qualm as bonuses to be seized when opportunity is presented.

Those are just for starters. Black bears love to claw apart rotten stumps and logs and lick up the ants swarming from nests inside them, taking the ant eggs as gourmet tidbits and never overlooking beetles and grubs. If a log is flipped over and a small salamander wriggles, it is also snapped up. Frogs, snakes, turtles, fish alive or dead, the raccoon that is after them if it can be grabbed, anything that crawls, swims, runs, or flies is fair game for the black bear.

Curiously, however, all these items, which might be lumped into the "meat" category, are actually only a small part of the average black bear's diet. It must be presumed that probably hundreds of thousands of years ago these carnivores were unable to supply their food requirements solely as predators. Even the teeth, which are thought to have evolved from a quite different arrangement long ago, are so designed that the two large pairs of canines for

*This Minnesota black bear is scratching for ants and insect larvae in a rotting deadfall. Bears relish these snacks.*

seizing and tearing are complemented by flat-topped molars at the rear of the jaws that serve as grinders for vegetation and other soft forage.

Thus the black bear and its larger grizzly relatives, large creatures that require a high daily intake to keep their big engines running, slowly adapted eons ago to omnivorous foraging. At least the ones that survived did. And over the broad expanse of geologic time, the animal that succeeded in expanding its range over almost the entire continent was able to do so because it developed a physical system and metabolism that could be fired by any forage available.

It is interesting that the large true predators—those that evolved on another tangent and were geared to meat only—have been the first creatures of the earth to decline drastically, and they often were and still are animals of rather restricted ranges. The African lion did well for centuries because forage animals of large size were always abundantly available and it was able to kill them with ease. But it could not spread its range to places where such forage was meager. The gray wolf of North America, at one time present over a vast range, could not sustain a high population level when its living forage declined because of human intrusion, and it could not live with man because it was too destructive to stock. The black bear, conversely, although badly affected by encroachment of civilization, still can eke out a living on the fringes because it can live on and will eat practically anything. The lesson in species survival is intriguing.

The black bear more than any other North American animal seems to love sweets. It raids bee trees in the wild, plunging right in, often until face and shoulders are covered with angry bees, but still the bear continues to gorge on honey until it is sticky from jowls to elbows. Now and then it also eats the bees. Blacks cannot resist fruit, and wild fruits in season are a mainstay of diet. Wild black cherries, chokecherries, and pin cherries are favorites.

It is common in bear country in New England and the Great Lakes region to find cherry trees with limbs torn off, and masses of bear dung full of pits below them. Chokecherry and wild plum thickets in the west get the same treatment. Hillsides everywhere that blueberries grow, and in some wooded lowlands the now rare huckleberry swamps, are literally combed by black bears.

In the old logging country of New England and the Great Lakes, cut over a hundred years ago, there are many places where apple trees either were planted near log camps or grew from discarded cores. These, and apple trees on abandoned farms, are favorite feeding places in late summer and early fall. Commonly a bear climbs a tree, even in somebody's backyard in the western mountains or back-country New England, and rips the tree all to pieces as it carelessly gorges with apples, and knocks more to the ground where it can get them later. Needless to say, these bears, if caught at their orchard raiding, don't last long.

Vegetable gardens and some small crop fields in bear country also take a drubbing. Patches of sweet corn are totally destroyed in one night unless the owner stays alert with his gun. Melons are a delight to any bear that lives nearby. So are tomatoes, pumpkins and even row vegetables. So also are fields of barley or green wheat plowed near forest edges. There is by no means the problem with bear raids on crops that there was in frontier days and even early in this century, simply because there aren't as many bears, nor as many crops located on the fringes of bear territory.

Wild nut crops are avidly enjoyed by black bears. These are, in fact, of inestimable value to the animals in fall because they are rich in fats and oils. Acorns, beech nuts, and in parts of the west piñon nuts are staples that help the bears put on thick layers of fat before denning. Not every year is a good acorn year, and some varieties of oaks bear only every two years. Because oaks of one kind or another grow almost everywhere within black bear range south of Canada, they are extremely important to bear survival. In off years when the acorn crop is low, blacks have to work harder for a living, and are consequently sometimes irritable and more likely to get in trouble with human neighbors.

Black bears are also insatiable bark eaters, eagerly devouring the inner bark of a number of trees, among them some hardwoods, such as maples. However, the conifers, particularly pines, get most attention. Timber companies, in the west particularly, look upon the black bear as an exasperating nuisance and expensive pest because at times it destroys thousands of young trees in new plantings. Varied tubers, of sedges, cattails, rushes, and numerous other plants, are dug for food, particularly in spring. Also in spring and early summer great quantities of green grasses of endless variety are consumed. When first out of hibernation, in fact, black bears in some places eagerly graze new green grass. In Ontario, for example, hunters sometimes quietly walk old log trails during a spring hunt. In these openings new grass shoots up quickly and bears soon find it.

During spawning runs of fish, black bears go after them with concentration. However, they are by no means as expert anglers as the grizzlies. In southern Canada and the lower United States they hang around creeks and larger streams when suckers and carp run in quantity. They are sometimes able to grab quantities of these near stream obstructions or where a small creek is packed with them. Hunters have high success by scooping up spring-run suckers and piling them on a bank for bait. Blacks are also hunted along the Alaska-coast salmon streams much as the browns are. But for the most part the black bears take their fish, such as spawned-out salmon, dead or dying, and others mostly trapped in shallows, where they are easy to catch.

It would be nearly impossible to list every-thing that goes into a black bear's stomach. It will eat almost anything, dead or alive, in the entire plant and animal world. This is really the secret of the animal's success on this continent. Seldom does a black bear simply go hunting for a specific kind of meal. It shuffles along, nosing here, nosing there, digging, clawing, ripping, overturning, sniffing, accepting anything that it finds. On occasion the finding is easy, as a whole hillside blue with berries, or a big moose turned to carrion by a hard winter. If no such bonanza is at hand, the bear keeps moving, picking a little of this, a little of that, endlessly stoking its nonselective stomach.

## Movements

Most movements of the black bear are directly related to its daily search for food. If there is an abundance, a bear may stay within as little as a square mile of range, but only as the abundance lasts. Much of the time its wanderings will encompass at least 5 or 6 square miles. In times of scarcity at least double that may be necessary. In addition, black bears apparently like to travel and wander, poking here and there much like curious travelers. There is no migration as such. A black bear may go to slopes where blueberries are ripe, even several miles from its spring and early-summer home. Or in spring it may follow low places where quick-growing plants of which it is fond, such as skunk cabbage, emerge in profusion. A beech grove within a mixed forest may draw bears from some distance when mast is abundant. But there is no definte winter and summer forage or definite seasonal movement over any appreciable distance.

Old males are inclinded to travel more than females and cubs. They're drifters who make the rounds, but meanwhile stay away from other bears except during mating time. Occasional fights between males occur, but no

Because of their short, closely curved claws that grip bark well, blacks are excellent tree climbers. When pressed to escape, they can shinny up a tall tree seemingly at a run. This species has been known to feed on fruit and nuts in trees and even doze on branches.

individual is eager to brace another unless it is extremely hungry, or hurt and irritable. Nor does the black bear have a strong home feeling. It may live out its life, if food is ample, within a short distance of where it was born. Or it may wander off many miles for no good reason and never return to its original home range.

Now and then a bear shows up in a location where none has been known for years, and hangs around. Just why, no one knows, and probably the bear doesn't either. Hundreds of bears have been live-trapped and moved, in parks and in settled areas where they have become nuisances. There are few instances of them returning, even when released only 20 miles or so distant, although very occasionally they have apparently felt strongly enough about "home," to have returned from long distances.

Blacks are strong swimmers. There are untold instances of them swimming in large lakes for several miles to reach an island seen from shore. Why, only the bear knows. A cabin built on an island in a large back-bush Canadian lake, where the owner felt sure no bear would ever bother him, may have its door smashed in or its windows demolished and the interior vandalized by a bear from the mainland that went looking, as black bears will.

Blacks are also excellent climbers. Grizzlies, except some very young individuals, cannot climb trees because their claws are so long and straight. Possibly their bulk also is a handicap. A black bear has shorter, closely curved claws that cling to bark expertly. An adult black, when pressed, can shinny up a tree almost at a run. Blacks also climb trees casually, after fruit or nuts, and sometimes probably just because they enjoy it. They have been observed up in a big tree dozing in a broad crotch, paws dangling, and at other times simply looking around as if pleased with the scenery. Coming down a tree, the bear has to move rear end first, sliding and grabbing with its claws. While still a few feet above ground, it lets go and plops down. Very occasionally a treed bear will jump, anywhere from 15 to 30 feet, landing rather awkwardly but without apparent harm.

Much has been written about whether blacks move chiefly by day or by night. A common belief, repeated over and over, is that they are by nature diurnal but when harassed by humans become nocturnal. This is mostly nonsense. A black bear moves whenever it pleases. In campgrounds panhandling bears rummage in the garbage buckets at midday and at midnight. They may be observed pottering along a stream course by day, or seen on jeep trails far back in the wilderness areas in the middle of the night.

It may be that bears on the fringe of settlement switch to almost complete nocturnal habits as a measure of furtiveness. However, black bears taken by hunters out of fly-in spike camps far back in the Canadian bush, in places where they've never seen a human or smelled one, and bears far back in the mountains of the west, living in total seclusion, habitually come out very late in the afternoon just before dusk to feed on a pile of carrion left from a winter-killed elk or deer. The last hour of daylight is invariably the best hour of hunting. Conversely, the same bear may be found fiddling around on a slope at midday—if it happens to feel like it.

The gaits of the black bear are several. Undisturbed, it ambles along flatfooted in a rolling, almost aimless manner. It also walks swiftly, head swinging, at times, and when in a hurry or frightened it breaks into a swift gallop. Going all-out, which it can do for short bursts, it may reach 25 or 30 miles per hour. Its slower gallop covers much ground, and can be maintained for much longer distances. In front of a dog pack, a big bear with stomach not overloaded can run for hours, doubling and circling, pausing to whack a dog here or there and then moving on. Some never are treed or bayed before wearing out the dogs.

After feeding full, the black bear makes a bed and dozes several hours. The bed may be in a thicket of brush, under a blowdown, or just anywhere that is comfortable and secluded. It does not habitually use the same bed more than once. Occasionally a bed will be a mound of leaves or grass scraped together for greater comfort, but most of the time the animal is not very particular. Some of the more interesting bedding places of black bears are those out in the open. They have been seen lying sound asleep on a rock ledge that overlooks a valley where they have been gorging on carrion or dead and dying fish in a stream. It is also not unusual to see a bear lying in the sun high on a mountain slope, even up atop a big flat rock spang in the open, forelegs dangling over the edges.

In areas where they have established goals, such as a slope where they sleep and a valley where they feed, or along a stream course traversed often, they wear broad, distinct trails from point to point. These are seldom as common or deep-rutted as those made by the big Alaska browns and some grizzlies. Probably the most active time for the black bear is when it is first feeding in spring and must hunt hard to fill up, and late in the fall when it is constantly gorging and laying on fat before entering its winter den.

## Denning

The amount of fat a bear puts on, when food is abundant, to see it through the long winter sleep is phenomenal. An extra 100 pounds is common, and twice that in northern areas is not especially rare. Many wildlife enthusiasts, and hunters, are under the impression that the black bear, and grizzlies, hibernate. They don't. When a bear dens up for several winter months, there is no drastic change in its metabolism. It becomes drowsy, and goes to sleep.

Much is still to be learned about the precise physical aspects, and what triggers the sleepiness. Temperature certainly is a factor. Southern black bears, undoubtedly with the instincts dating from the early evolution of the species, may go into a den for only a week or so, or only a few days, or intermittently during winter. Some may be less active but not den up at all. In far-northern regions black bears may enter dens as early as mid-October and stay denned for five months or more, emerging in April or as late as early May. In more moderate winter climates the den stay is shorter, from about mid-November until March or early April, or even less. Conceivably the urge to retire to a comfortable, warm den and sleep through the severe weather is an instinct toward survival acquired long ago. Winter is the time of scarce forage. The energy required to keep alive—if enough food for keeping alive were available during the difficult winter season—might defeat the purpose. The bear would die because energy expended in cold while gathering a livelihood would overbalance the intake of food.

By putting on fat in the fall, the bear is able to go to sleep and wait out the difficult time. Its body temperature drops, but not much. Its breathing rate is only slightly slowed. It goes to sleep, but is never in a state from which awakening is difficult. During warm spells, bears may come out of dens, groggily wander about, then go back. When the first of the drowsy season comes on, there is an interesting sequence the bear follows quite instinctively. It apparently feels too dull and drowsy to eat for a few days. During that short fasting period the entire digestive system is of course emptied.

However, it would hardly do for the stomach and all of the intestines to be flatly deflated for so long a period. Thus the bear takes on some roughage as padding, commonly needles of conifers if they are available, dead grass, or fallen leaves. A substantial portion of this material passes through the stomach and intestines and lodges in the lower bowel, where it forms a firm stopper of sorts. If the bear entered its den with stomach gorged and intestines full and went to sleep, masses of excrement would be piled inside the den. The swiftly shriveling stomach and intestines, with the lower bowel stopped up, avoid this.

Dens are selected in a wide variety of locations. A cave serves well, or a hollow beneath a blowdown. A hole under a cut bank may be chosen. Some blacks dig dens in the side of a slope, under tree roots, or among rocks. Still others, less finicky, may simply curl up beneath a low-hanging evergreen and let snow cover the branches, or even make a bed in dense brush or high dead grass. Females use more care in choice of den site as a rule than males, presumably because it is in the winter den that the young arrive.

When the animals first emerge from dens in spring, they have lost some of their fall fat, which as been burned up during the long sleep. They drink water and eat soft foods. Fresh sprouting green grass, full of liquid, is often taken in large quantity right at first. Its juices, plus water if easily available, apparently soak the contracted, empty digestive system and purge it, getting it in working order again. By the time the bodily functions are geared up once more the bears are feeding heavily, and now the urgings of the mating season are at hand.

## Breeding

Only during early summer each year do males and females consort. The period during which females are ready to be bred begins as a rule in June. Not all allow attentiveness of males at the same time, and so the breeding season lasts sometimes through all of July. During that period one male may stay for several weeks with the same female. This may be the only female he finds. Or after a short pe-

riod the male may wander off to look elsewhere, thus servicing more than one sow.

Black bears may be crotchety and antisocial at other times of year, but during the mating season the boar is an ardent swain, nuzzling and licking the female, following her constantly, feeding sometimes nose to nose. The two may rear erect and embrace. They paw each other and occasionally wrestle playfully and seemingly with great affection.

Females are fully adult before they are able to mate. A female cub born in a den in February or March of one year will not breed until she is in her third summer. Then, possibly because young bears need parental care longer than many wild animals, she skips at least one year, coming into heat again two years later when she is free of her youngsters and they are on their own.

The growth of embryos in bears is quite different from the system among most other mammals. After fertilization, the female's egg cells do not begin immediate development, and do not attach themselves to the uterus until several months later, in fall. Just how this system evolved, and why, no one is certain. Bear cubs are extremely tiny at birth, and helpless—almost, one might say, premature. Delayed implantation (embryo development) may be to assure that the female is denned up cozy and warm before the cubs arrive, so that they are not brought into a bitterly cold world where they could not survive.

It may also be that the habit of denning is matched to the season where food is scarce, as has been noted. Youngsters could not possibly learn to take solid food in dead of winter. Thus with the heavily fattened mother giving warm milk in a weatherproof den, both she and the young are able to wait out the winter and the coming of spring forage. Man may believe he could have planned all this better and simpler—as by having mating season in fall and obviating the need for latent embryo development. Be that as it may, the unique ar-

*Female black bears, or sows, in suitable range have one or two cubs every other year. Occasionally triplets are born. Newborn cubs are blind and almost helpless during winter in an underground den. They weigh less than a pound.*

rangement seems to have worked well, and with high success, for bears over many centuries, and the "why" will always be only supposition.

## Birth and Development

Baby black bears are almost unbelievably small. It seems preposterous that 350-pound mother bear should give birth to young that can be held two at a time in the hand of an average man. Each weighs from 6 to 8 ounces. Many females giving birth for the first time produce only a single cub. If so, from then on twins are the rule, and triplets not rare. Seldom

are there more, but regular litters of four to six have been known.

The litters are born with eyes snugly shut. The cubs are covered with very fine hair so short that they appear naked, but in the warm den they have no trouble worming down into their mother's long, dense fur. Whether or not the mother is deeply asleep during the birth no one is certain. Probably she is only drowsy, is at least aware of the birth, and is able to clean up the offspring and snuggle them to her. One of the most intriguing physical attributes of the mother is the placement of her teats. Two are located between her hind legs, and four more are on her chest. Is this arrangement a safeguard to make certain the tiny cubs, born to a less than wide-awake mother, find milk, or are some of the extras standbys in case a whole litter of small bears occurs? Such are the puzzles of wildlife.

Most cubs are born during January and February. About six weeks later, give or take a few days, the youngsters have their eyes open and are beginning to be a bit frisky, but of course so far their only knowledge is of their mother and the interior of the den. A couple of weeks later, still unsteady on their legs but weighing ten times what they did at birth, they follow their mother from the den and begin to learn about the outside world.

Black bear cubs are often an exasperation to their mother. They are like bubbly, exuberant small children, forever teasing and chasing and rough-housing each other, climbing, and tumbling into everything, and bedeviling their mother. Now and then the mother, though exceedingly protective in face of any danger, and always suspicious of disturbance or the presence of another bear, finds it necessary to bat a cub with a swipe that sends it flying and crying.

The small bears continue to nurse, but soon are picking at any food they see their mother eat, and sharing with her any kill or forage discovery she makes. They climb trees agilely now. In fact, at sign of imminent danger—or what the mother suspects may be danger—she herds them together and sends them scurrying up a tall tree. Boar bears that happen to pass by have no compunction about killing a cub. But woe to one that tries it. The mother flies into a violent rage and will fight instantly.

*This is a unique photo of a black bear looking up into a tree at the photographer sitting in a blind near a bear bait. Although most black bears are not naturally aggressive toward man, any black can be dangerous. This one was not.*

By late fall the youngsters begin to look more like replicas of adults. They are weaned but loath to leave their mother. At denning time, they go in with her, and all sleep together congenially. When spring arrives and all leave the den, family ties begin to fray. Very occasionally young bears may stay on into the fall with their mother, but the majority begin lives of their own during that second summer.

## Senses

The most highly developed sense of the black bear is smell. Like almost all creatures that live in a habitat that has much dense cover, it is important that the bear be able to pick up scents at long distance. A keen sense of smell is also important in locating numerous varieties of food. A black bear zeroes in on carrion from a half-mile away when the breeze is right. It can locate abundant ripe fruit the same way, and easily select by smell the types of vegetation that make up the bulk of its diet.

Hearing is also good. Many nature writers equate keenness of hearing with ear size. By this measure, a mule deer would be able to hear better than a whitetail deer, whereas probably their hearing abilities are about equal. Nonetheless, among bears, the black

does have the largest ears, and it does have extremely sharp hearing. Again, this sense undoubtedly has developed to a high level because the animal lives in and hides in heavy cover. A bear must always be listening for disturbing sounds. The furtive and secretive black bear, bringing both senses of scent and hearing to fine tuning, is well aware of what is constantly happening for several hundred yards around it, and thus can drift off unnoticed when its suspicions are aroused.

Its eyesight is not very good. Possibly its small eyes are nearsighted. It sees movement readily, but does not quickly identify motionless objects unless it can smell them. In its daily foraging, extra-sharp eyesight over distances is not any great advantage, and in cover it cannot see far anyway.

## Sign

In a forest where the ground is carpeted with pine needles or other matter that does not readily accept imprints, a black bear moves silently without leaving any sign at all. But a bear that lives in and feeds over a specific domain invariably marks its presence plainly. Tracks show in mud, dust, or snow. The bear walks flatfooted. It is the hind foot that leaves the print rather similar to that of a human barefoot track, except that the claw marks show in front of the toes and the little toe and big toe are transposed in placement compared to the human foot. The print of the front foot shows the pad or ball of the foot, plus toes and claw marks, but may or may not show the small, rounded heel.

The only track with which a black bear print might be confused is that of the grizzly. However, among tracks of adults the grizzly is far larger, and the stride is longer. Tracks of adult blacks of average size are as follows: forefoot, not counting the heel, about 4 ½ inches long by 3 ½ to 4 inches wide; hind foot, overall, 3 ½ to 4 inches wide by 7 inches long.

Walking or bounding, the hind-foot prints are ahead of those of the fore feet.

One of the most interesting "track" signs of the black bear occurs on the bark of a tree here and there, of species that have fairly smooth, soft outer bark, such as poplar and birch, maple, aspen, and the smoother-barked nut bearing trees, for example, beech. When a bear climbs one of these trees, each claw sinks into the bark. The wound bleeds but eventually heals and in time leaves a scar, usually black or at least dark on lighter-hued bark. At first glance these old bark scars, which with age become rounded and broadened, look exactly as if they were imprints of the toes. Of course all they mean is that a black bear was there sometime, several seasons previously. But if such scars are numerous they may indicate that this is perennial bear range and that a search for fresher sign is worthwhile.

Blacks also rake trees with their claws. Rough-barked evergreens commonly have claw marks, fresh or old. Bears may make them when climbing, or they may simply rear up and claw a tree. They also rub against certain trees, especially when shedding, and wear the bark smooth. These trees may also have claw marks. Torn bark where bears have ripped it open to lick juices or eat the soft inner bark, shredded rotten stumps or logs where they've sought ants and beetles, broken limbs on wild fruit trees such as wild cherry—all mark where black bears have been foraging.

Any sharp-eyed woodsman is ever alert to all these signs. He also watches for bear trails. They appear only in special places—along a stream, perhaps, or in a habitually used bedding place near a feeding area and the abundant food. Bears coming to garbage dumps or carrion, perhaps a dead big-game animal, where they feed daily for some time, make well-packed trails that are easily seen. Hunters look for copious droppings. It is easy to judge from these whether the bear has been using a place regularly, or recently. Soft foods—fruit and grass—form manure that is also soft and left in large piles. More solid foods, as meat, tend to make droppings firm, 1 to 1 ¾ inches in diameter and in pieces several inches long.

## Hunting

Every hunter always gets excited when he spots fresh bear sign during an open season. But most of the time, aside from relaying the message that a bear lives here or recently went through the area, it is not likely to pay off with a rug on the floor. It is estimated that at least 90 percent of all black bears annually bagged are taken as incidentals by hunters after other big game when bear season also happens to be open. Deer hunters, because they are the most numerous and because deer and black bear commonly use the same ranges, collect most of them. Thus, most of the hunting is not done "on purpose."

There are two chief methods of purposely hunting black bears. One is with dogs, the other is with bait. Tough, highly specialized big hounds are mandatory for bear hunting, and several are needed. There's no denying that hunting with hounds is a most dramatic, and also exceedingly rugged, sport. But because a trained pack is the basis for this endeavor, plus location in an expanse of habitat that has a large enough bear population to make it worthwhile, the only sportsmen who ordinarily pursue this hunting are the dog owners and their friends, or those who book hunts with an outfitter who specializes in running bears with dogs.

In some states no dog hunting is allowed. Where it is legal, obviously there are not very many dog packs. It's expensive to train and keep the dogs. It also takes a lot of one's time, and quite a few are killed by bayed quarry. There are, however, a number of outfitters, particularly in the west, who specialize in this hunting. It is fairly expensive, and also chancy. Here and there a booker advertises guaranteed hunts. These may or may not be legitimate. A no-bear-no-pay guarantee is usually all right. The outfitter gambles, knowing he has good dogs and a good territory. But there have been a lot of shenanigans over late years in the bear-hunt-with-dogs game. A booker who guarantees a bear, period, is one to stay shy of. In some of these cases bears live-trapped or bought surplus from zoos and kept in pens are released just before the hunt. However, a legitimate run with dogs is sporting indeed. It is not necessary to explain the technique here, because it is invariably done with a guide and pack.

Hunting over bait is one of the most successful methods of black bear hunting. This is done almost entirely in the spring, as soon as the hungry bears have come out of their dens. Here again, most bait hunts are booked with an outfitter. He puts out or finds a bait, perhaps a dead big-game animal that died during winter, an old horse or cow destroyed and placed in a location where bears will find it, or maybe a pile of freshly netted suckers from a stream. At least one astute guide is known to use gallons of cheap strawberry jam, dumped and smeared over a crisscross pile of down timber. It is not as expensive as buying an old decrepit cow, it is easily transported to a good site, and as it dribbles here and there it makes the bears work over a number of visits to get it all.

Garbage from camps—for example, winter lumber camps in the Canadian bush—also serves for bait. Many winter logging camps may house dozens of men and work horses all winter. They are abandoned before ice-out. Guides often take hunters to these camps, for bears hang around them after coming out of dens, eating grass that sprouts and leftover grain, licking salt, and digging in the dump.

Baiting, or hunting over a natural bait, is also not legal everywhere, and for that matter, there are spring seasons in only a few states and provinces. Still-hunting along old log roads in spring is sometimes successful. So in fall is riding horseback in mountains, glassing for bears, or prowling along stream courses.

*Since backcountry black bears usually avoid man, hunters often use hounds trained to find bears and bring them to bay.*

But still-hunting for black bears, especially in fall, is a long gamble. Very occasionally a stand beside a well-worn bear trail may bring good fortune. Along the Alaska coast blacks are hunted the same as browns (see the chapter on grizzly bears) along coastal rivers or from a boat cruising the shoreline.

Even in territory where black bears are plentiful, purposely hunting one without dogs or bait is never very productive. They are simply too shy, and a man cannot cover enough ground to be at the right spot at precisely the right time except by total chance. Even when hunted with dogs, and in some instances with a score of hunters spread out taking stands where a jumped bear may pass within range, the gamble is still long. Many a bear offers an exciting run, but is never seen. In addition, many a good dog has been killed, sometimes half of the pack if they are tenacious and have bayed a bear that won't tree, before hunters can get to the racket and dispatch the quarry.

Nonetheless the black bear is a prime and exciting game animal. Happily, even though black bear range is only a remnant of what it was originally, and the number of bears likewise, this animal is in no danger, nor is it likely to be in the foreseeable future. It is one of the most interesting of all North American game animals. For those who would hunt it, or photograph it, or simply try to observe it in the wild, it also deserves a healthy respect. A clown the black bear may be, but ever a totally unpredictable one.

# Mountain Lion

*Felis concolor*

*The cougar, mountain lion, puma, painter, or catamount is by any name a sleek and handsome, tawny cat and a skilful predator. Once it ranged almost across America, but its final stronghold is in the remote western wilderness. A few of the cats barely survive in southern Florida and possibly in eastern Canada. They are rarely seen except by hunters.*

"**M**ountain lion" is the name by which this animal, the largest unspotted, long-tailed cat of North America, is known today to most people. It is not an especially apt name. It has come into common modern usage chiefly because most of these cats still left on the continent are found in the large, wild forested expanses of the western mountains. Originally, however, colonists and later settlers and explorers in widely separated locations might not have known what animal others referred to in chronicles of the frontier that were filled with tales and legends concerning this big cat. It was known by numerous names—cougar, puma, panther and its corruption "painter," catamount, and many others.

Local names galore were born from the vast original range over which the creature had spread and the astonishing variety of habitats to which it had been able to adapt in the process. The first written account of what is presumed to have been the cougar was by Columbus, telling of sightings in coastal Central America. Coronado's 1540 expedition in the southwest also notes the animal. Undoubtedly some of the first of the big cats seen by white man in present-day U.S. territory were prowling the swamps of Florida and the deserts and mountains of the west. When explorers of varied origin first touched land on the Pacific coast, they discovered the mountain lion. The same was true in eastern Canada.

Indeed, wherever pioneers pushed into new territory, from all of the east coast to all of the west and from what is now southern Canada to the bottom of Mexico, the tawny cat was present. Mountains, deep swamps, wooded stream courses meandering across plains, deserts of dense cactus and thornbush all were home to it. In fact, this enormous range over all but the northernmost quarter of the entire North American continent, wherever ground cover grew for it to hide in and hunt in, was only a beginning. Known by scores of differing names in English, Spanish, Indian, and even French, *Felis concolor,* "The cat of a single color"—without mottlings, stripes, or spots— had prowled its way down through the centuries clear to the southern tip of South America. Of all land-dwelling mammals, it had colonized and long held the greatest expanse of territory.

In North America particularly, it was not able to hold its vast possessions long. A creature of the truly wild places, in need of large hunting areas and exceedingly shy of disturbance, it could not tolerate the presence and competition of the white man. Conversely, neither could the newly arrived white man tolerate in his intimate domain a predator so large and so deft at the hunt and the stalk. As settlement inexorably reduced lion habitat, as settlers competed with the lion for its natural food, the deer and similar large forage animals, the cougar was driven back, yet was still ever on the fringes of human domain. As game dwindled, easily killed livestock was placed in ever growing numbers at its disposal. Quickly it acquired a taste for horses in particular, and for sheep, pigs, and calves simply because they were available and easy to catch.

Because of the pressure brought to bear on them by shoot-on-sight settlers, by government trappers, and by the swift shrinking of suitable habitat, the lion was gone from most of its eastern haunts by the turn of the last century. In various places lions had been bountied since early Spanish days. From the beginning of the cattle business cattlemen had instituted bounties throughout the west. States continued them. Yet there was still enough western U.S. wilderness and more in Mexico to sustain a moderate unendangered population. Within recent years the mountain lion has received much attention, and protection, from game-management people. In almost every state, bounties have been removed and it has been placed on the game-animal list. This allows regulation of hunting as to method, time of year, and number that can be legally taken, and also allows total year-round protection in states where the animal's existence appears precarious.

Officially today the mountain lion is considered to range in suitable wild habitat over much of British Columbia, southern Alberta, parts of western Saskatchewan, southward throughout the Rockies and the Pacific Coast states, border to border across southern Texas, spottily in the southern swampy expanses of the Gulf States, and in southern Florida, as well as over much of Mexico. However, the official range doesn't tell the most interesting part of the modern mountain lion story.

Periodically sightings are made that appear well authenticated in widely separated places from which the lion was presumably long ago extirpated. Reports come out of Canada's Maritime Provinces. Others originate periodically in Maine and other New England states. Massachusetts has officially pursued several scattered reports over past years, and game biologists are convinced the few sightings and tracks have been authentic. New York State, the Appalachian region, and isolated locations in the Ozarks have all reported the presence of an occasional mountain lion.

This big cat, one might think, could not

## THE MOUNTAIN LION

**Color:** Varying shades of light brown, from yellowish to reddish; in some areas grayish; very occasionally melanistic (black); without spots; face usually darker than body color, on sides of muzzle, nose bridge and forehead, but with front of muzzle white to grayish; underparts graywhite; whiskers white, eyes yellow, tip of tail darker than body, to black, geographical races differ in richness of paleness of overall coloring; seasonal differences also, richer in summer, grayer in winter.

**Measurements, adult males:** Overall length 7 to 8½ feet including tail, rarely to 9 feet, the tail 2½ to 3 feet of the total length; height at shoulder 2 to 2½ feet or occasionally slightly more.

**Weight, adult males:** 135 to 175 pounds average, exceptional specimens over 200 to a maximum of 275-plus.

**Females:** 35 to 40 percent smaller.

**General attributes:** Solid, unspotted color, long tail, large size; slender build with comparatively small head; ears without tufts as in the smaller, short-tailed lynx and bobcat; loose-appearing skin of belly; short, smooth fur; shy, secretive personality; very graceful movements.

*Range of the Mountain Lion*

possibly reside near human habitation without its presence immediately becoming known. But it is superbly equipped to lead the secretive existence, moving on padded feet without sound, sifting like smoke through the heaviest thickets and forests available, hunting by night and lying up by day in rocky crevices or on slopes dense with brush, ever attuned with astonishingly sharp senses to the slightest indication of man's near presence.

To illustrate the uncanny ability of an animal this size in keeping its local existence unknown in suitable circumstances even on the fringe of large urban areas, an incident in Texas is a classic example. Careful studies of the lion in Texas had previously indicated that the animal was endangered there, with a probable total population of no more than fifty animals. These were in the arid southern so-called Brush Country near the Mexican border, and in the neo-frontier Big Bend Country of far-western Texas.

The central-Texas Hill Country 200 miles north of the border traditionally is a mohair and sheep area. Large predators such as coyotes and bobcats are not tolerated and are rarely found. Lions have been unknown there for decades. Yet that year lion tracks were discovered on a ranch of modest size but with dense cover, within a few miles of San Antonio,

*Any cougar blends quietly, imperceptibly into its home country. Adult males weigh 135 to 175 pounds, with the rare one exceeding 200. Females are 35 to 40 percent smaller.*

which with environs totaled over a million people. Lion sightings had been reported for several years in the ranch vicinity, but were scoffed at. Eventually, however, residents were astonished when two adult male lions were trapped. In 1975 another big lion was shot by a deer hunter in another part of the Hill Country where lions are unknown, and in 1976 one was struck by a car at night near the village of Ingram, where mountain lions have been unknown since frontier days.

Thus no one can say with absolute certainty precisely where mountain lions still range. Doubtless the places where they still exhibit well authenticated populations and are occasionally and successfully hunted as game trophies form the important islands of population. Various estimates have been made as to how many mountain lions still exist north of Mexico. Numbers have been set at 200 or 300 east of the Mississippi, 5000 or 6000 west of it. These are strictly guesses. Counting mountain lions is like counting mirages.

By estimating the hunting domain of a lion, dividing the figure into the suitable square-mile habitat in a state, and then applying figures of bountied kills or sport-hunting kills, a rough estimate can be made of how many lions may be—or *might* be—present. However, there are scores of outdoorsmen who have spent a lifetime in sign-spattered cougar country without ever glimpsing one alive and free in the wild. Thus population estimates, though perhaps helpful in management, are just that—estimates. Some wildlife-management specialists suspect that the cougar is slowly making a minor comeback in certain sectors of the east, perhaps partially adjusting to life on the wilder fringes of human settlement.

Numerous geographical races of mountain lions developed over the centuries because of their vast range. Such differing populations usually evolve because of isolation from others, in a more or less specialized habitat. At least thirty have been named, split about evenly between North and South America.

Many of the subspecies are based mainly on size and color differences, with minor physiological differentiations. There is little reason for the layman to be concerned with them. The mountain lion, cougar, or puma is for any practical purpose the same animal wherever it is found. In general the largest individuals have evolved in the extremes of the range, both north and south, that is, in colder climates of the north in the United States and Canada, and at the far-southern portion of South America. Mountain lions of the tropics are decidedly smaller.

Throughout the long history of the relationship of people and pumas, from the times of ancient Indians even to the present, so much legend, ritual, and yarn spinning has accrued relating to these cats that it is difficult to separate truth from fiction. Added to the puzzle is the true mystery of the cats themselves. Everyone, it seems knows *about* the beast—but those few who know the most about it admit that they actually know very little. Probably no large animal has done a more thorough job, by its retiring habits, of fending off familiarity.

Indians of various tribes on both continents anciently revered the lion, used its hide for very special leather, wore and made decorations and ceremonial icons from skulls, teeth, hides, and claws. Ancient Incas used thousands of their slaves in hunting drives to gather abundant pumas. Many native peoples ate and relished the mountain lion. In fact, within recent times—even as late as the 1940s—backcountry "crackers" in Florida would flock to a southern swamp village when word was out that a lion had been killed, hoping to get a piece of the meat, which was considered a delicacy. In scattered locations in Mexico natives have always watched closely for circling buzzards that might indicate a lion kill, so that they might attempt to rob the cache left after the predator had eaten its fill.

Tales of mountain lions attacking humans and trailing woodsmen, hunters, and mountaineers for miles are legion. Are they true? Some are. Numerous trailing instances have been well authenticated, but conclusions have invariably been that curiosity, or some unknown attraction of following a trail, was the reason. It is doubtful that a predatory urge was ever the motivation. Nonetheless, there are a very few well-documented cases of attack without provocation.

The all-time classic, presumably, was the killing and partial devouring of a boy years ago by a cougar in Washington. As recently as the mid-1970s a New Mexico resident was mauled during an apparently unprovoked attack. A few provoked attacks have been recorded. What may have happened in colonial times is now impossible accurately to assess, but it is believed that in more recent times, within the past century, there have been possibly two dozen people killed by cougars.

Regardless, the big cat cannot be considered truly a dangerous animal. Hundreds have been hunted with dogs, treed, and approached without danger. Many of these have been roped alive by an experienced hunter who climbed up after them. In rare instances, possibly, a lion may make an error of judgment, believing a human to be a forage animal. The bulk of experience with the cougar in all of its range indicates that it is an extremely shy wilderness character to which human scent is frightening, one that would prefer to run or even give up rather easily rather than fight man, its only serious enemy. That a creature so extraordinarily powerful should have with so few exceptions so successfully and intentionally avoided mankind for several centuries, while being crowded endlessly by human incursion, would indicate that the puma is basically a retiring and inoffensive predator.

Most Indians, who knew the big cats on the whole more intimately than we do, must have thought so. They greatly admired the hunting prowess of the mountain lion, its power and quick skill in killing adult deer and other large animals. Some of them prayed to it, hoping to have its hunting abilities bestowed upon themselves. Probably the most striking evidence of the reverence in which the cougar was held by many earlier peoples, and a measure of its importance in nature's scheme in their view, is still to be seen in Bandolier National Monument in New Mexico. It is a sculpture carved in native stone by ancient ancestors of modern Indians. It is believed to have been a shrine of sorts, of a pair of mountain lions standing together.

## Habitat

It is not possible to give a specific description of the mountain lion's habitat. Like all animals that have been able to adapt to highly varied conditions of terrain, vegetation, climate, latitude, and altitude, no one description of its usual surroundings fits the larger picture. Nonetheless, certain general characteristics are apparent in all situations. For example, the lion, like all cats, stalks its prey and then makes the kill with a brief, swift rush and pounce. It must therefore have cover in which to hide while hunting. Thus mountain forest, densely vegetated swamps, and the thornbrush and cactus of deserts all answer the purpose.

However, and obviously, proper forage must be present in ample quantity. Therefore where a puma lives is a bit like the puzzle about whether the hen or the egg comes first. Although the lion has been known to eat snails and catch mice, because of its size it must have available large prey animals. It is pertinent to note here the difference between predators with an omnivorous inclination, such as the bears, and the wholly predaceous creatures of which the mountain lion is a classic example. A black bear makes do nicely on a vegetarian diet when meat is not readily available. The lion must have meat, in large quantities. And it does not relish carrion. Thus, whereas the

*When not actually hunting, to which they must devote much time, mountain lions retreat to rest and sleep in very remote areas, often in caves or under deadfalls. Females have their cubs in such places.*

bear, a complete opportunist, may visit a dead and rotting elk until every consumable ounce is eaten, and then eat grass, the lion is rigged by nature to endlessly need fresh kills.

Only because large herbivores—deer especially—were able spread their range so broadly was the lion able to follow. Otherwise it could never have adapted to the swamps, the high mountains, and the deserts. Further, because large, antlered and horned prey animals require a rather large living area, and are restricted in numbers to what a range of given quality will support, the lion must roam widely in order to make a living. Even when

the prey is extremely abundant, deer perhaps overpopulous with as many as fifteen, for example, to a square mile, the predaceous lion cannot hope to successfully stalk and kill each one it sees.

So it might be said that the mountain lion can be anywhere that ample cover exists for its hunting, but with the qualification that it is restricted to places where prey animals of large size have colonized first. Because in general the large prey creatures require large wild expanses in order to forage and propagate successfully, the lion must have an even larger wild expanse in order to glean its living from them. Thus the mountain lion invariably is found most numerously where there are large, unbroken wilderness expanses affording ample cover for its secretive ways, and for a goodly supply of its prey.

To be sure, the habitat picture is drastically changed from what it was before settlement. Then lions might roam almost everywhere. Now they are forced, with some straggler exceptions, to remain in the few suitable true wilderness situations still remaining. However, as long as food and undisturbed prowling cover are available, the lion is not choosy about its general habitat type. Mountain lions have been found high above timberline, to as much as 13,000 feet altitude. The Everglades and Big Cypress swamps in Florida still sustain a modest population. The deserts of the southwest and Mexico are still havens.

In the Rockies, where the major cougar population is now confined, the big cats favor the most rugged and remote canyons. Hunters running dogs after them can well attest that the best lion country is invariably the most difficult of access. Forested areas slashed by steep cuts faced with jumbled rock slides, pinnacles, and ledges are typical lion habitat.

## Feeding

These regions are also typical habitat for mule deer and elk. Of the long list of prey animals found in the diet of the cougar throughout its North American range, deer are the most important. Undoubtedly this is because deer, both whitetails and mule deer, are the most numerous large animals present. It might seem to a hunter who doesn't always get the deer he goes after even with a long-range rifle that the wary creatures would present insoluble problems to the lion. Studies have shown that on good deer range, a lion

*A mountain lion, after stalking its prey on the ground, pounces for the kill.*

traveling its hunting circuit makes a kill at least once each week, sometimes more often. That is at least fifty deer per year. Certainly the lion, like all predators, is an opportunist. If an easy kill of some other animal is presented, hunger is sated then, and there. Regardless, the lion is a severe predator upon deer, and deer are constantly targets of its hunting efforts wherever they are available.

Nature writers of less than ample field experience, and many of the modern "instant ecologists" who have lately discovered "the environment," are fond of claiming that a predator such as the mountain lion, unlike the sportsman who selects the best trophy, takes only the weak, the sick, and the old, and thus is a boon to natural herd management. This is utter nonsense, as many studies of lions have proved, in some instances where researchers spent weeks living on the trails of their subjects and tabulating kills.

It is true that the lion will seize opportunity to make an easy kill. Certainly this will account for deer or other prey that happen to be partially incapacitated. But just as the big cat seizes these opportunities, it seizes any that are presented. Whatever deer a lion spots when it is hungry, in a situation where it senses that a successful stalk may be made, it makes the attempt. Some alert, vigorous deer in fine health are successful in escaping. Some aren't. Out of fifty deer taken in a year, it is probable that most are in good health. They are the ones most active in feeding and moving about and thus more often present themselves as targets of the stalk. Many are among younger deer, simply because they are somewhat less wary. Conversely, many are prime, trophy bucks that just happened to be victims because a mountain lion noticed them.

Inept nature writing and disgracefully faked movies have falsified the manner in which the lion makes its kills on large animals. More often than not, the story or film scene is of the cat lying hunched and tense on a limb high above a well-worn deer trail, or on a rock ledge beneath which deer pass. As the unaware prey moves into range, the lithe cat makes its spring. Flying through the air with forepaws outstretched, claws extended and fangs bared in a snarling mouth, the lion slams atop its prey, smashing it to the ground.

It is possible that some lions sometimes have happened to be lying on a branch or a ledge and have grabbed the chance to jump upon an unsuspecting deer. Lions often lie on ledges to rest. They seldom climb trees, however, unless after food up in the branches, such as a porcupine, or when chased by hunting hounds. A lion that habitually tried to waylay its quarry would inevitably starve to death. Almost all kills of large animals are made by a belly-to-the-ground stalk, after the creature has been located by scent or sight, or by zeroing in one or the other or both senses when a sound has alerted the cat.

During the stalk every bit of cover is utilized by the cat to cover its movements. It may keep moving parallel to a trail along which a mountain sheep or a deer is traveling, slowly narrowing the rush distance. Unlike the wild dogs, cats are not geared to long chases at high speed. The lion moves with astonishing agility and power in a short burst. But like the housecat stalking a bird on a lawn, it stays low, gets close, then in a violent and beautifully coordinated rush closes the distance and leaps. A lion has been observed moving in on feeding deer hunting as it invariably does, into the wind, in a mountain meadow where only foot-high tawny grass was available for cover. With the end of its long tail twitching, eyes burning, forepaws fully extended and head flat down between them, the cat crept along, freezing for long moments immobile, moving at intervals without the slightest sound. When within only a few yards it catapulted into action—the awesomely powerful, swift rush, the leap to land on the shoulders and neck of a deer.

The mountain lion is a brilliantly designed

*Virtually everywhere cougars live today, deer are the principal prey species. But cougars are beneficial predators and play a role in maintaining the health and vigor of deer herds by culling the weaker ones.*

killing machine. Like all cats, it has retractable claws. Sheathed in their soft packaging, they are never worn dull from contact with rocks or hard ground. They are curved like talons. When about to make the final rush on stalked prey, the lion gets all four feet well placed under it. The claws are now extended so that they get a turf grip to help hurl the muscular body ahead.

There are five toes on the front feet, four on the rear. However, the "thumb" toe and claw of the forefoot is of little practical use. When the kill leap is made, the hind feet are commonly used to rake the flanks and belly, but it is the forefeet that are most important. The claws on one foot sink for a grip across the shoulder, and the other set of claws rips into the deer's nose and head, in an instinctive attempt to pull the head around and backward. Meanwhile the great canine teeth set in unbelievably powerful jaws sink into the spinal region. Sometimes the combination of the downward bite and clench of the jaws, plus the twisting of the deer's head, breaks the animals neck instantly. Teeth may even slash the spine in two.

Of course the technique of each kill varies, according to the situation presented. An interesting physical development of the big cats—and smaller ones, too—is the unusual development of the bones of the shoulders, specifically the clavicle. Whereas the wild dogs are fashioned for running, seizing prey with teeth, and ripping at it, the cats are designed across the shoulders for immense striking and gripping power. The strike of a big lion at the end of its leap often involves such force of impact that a deer, not expecting it or braced against it, is knocked over.

By no means every stalk ends successfully.

The quarry senses danger and flees. Now and then a lion gives up without any final rush, aware that the prey has been alerted and that it cannot succeed. Or a deer, occasionally badly mauled, breaks free and runs. The lion fails to catch it. Many times a lion goes hungry because it fails in its attempts to kill a large animal. Then it must grab whatever is handy or possible. Stomach analyses have shown rabbits, mice, varied small animals. Ravenous lions have even tried skunks and coyotes, though apparently not with relish.

Young of elk or moose are fairly easy kills. But apparently the cougar is not skittish of at-

*The mountain lion does not limit its diet to antlered prey. This raccoon would be a welcome meal, though it is capable of giving a young cat a few nicks before succumbing.*

tacking the largest of antlered game animals. Lions have been known to kill adult elk and moose. On occasions when opportunity is presented, wild turkeys furnish a feast. In a few locations, for example along the Mexican border in the southwest and in Mexico, southern Texas, and southern Arizona, where the javelina or peccary is fairly abundant, individual lions form a particular taste for them. This may be simply because the little desert pigs are available, and fairly easy to stalk and kill.

As mentioned earlier, domestic stock on occasion is turned into mountain lion food. The cats seem inordinately fond of pigs, once they get a taste, and of sheep on a range, probably because they are easy kills. The lion also is renowned as a horse and colt killer, habitually turning down a chance at cattle when horse are on the same range. However, even large steers are not too difficult for a lion to handle. Stock killers seldom last long. Ranchers or government trappers go after them with determination.

A seemingly favorite meal for the lion is the lowly porcupine. Again, this may be because it is easy to catch. Females are known to climb and knock porcupines out of trees to cubs waiting below. The quills present a danger that can be severe. Apparently lions learn to flip a porky over and rip open the soft, quillless belly. However, it is all but impossible for the cats to eat a porcupine without eating quills or getting them into lips, tongue, or parts of the mouth. Curiously, an abundance of quills, possibly softened to harmlessness in the digestive tract, are often found in lion scats. But now and than a lion gets its mouth full of quills that fester and finally kill it, either by infection or by starvation because the animal cannot eat.

Odd as it seems, the porcupine therefore may be considered one of the few potential enemies of the mountain lion. In isolated instances, a lion may also be an enemy of its own species. Males kill kittens now and then,

even eat them, and an old lion may kill and eat a smaller, younger lion that gets into its territory. In the tropics, jaguar and lions may cross paths and do battle, but the heavier jaguar can hardly be considered an important enemy, and may even be whipped by the more agile puma. Probably more often than one might believe a lion is injured severely during the attempt to kill a deer, elk, or other big-game animal. Injuries of this sort, by falls over ledges or by horn or antler damage to the attacking cat, are well authenticated.

Once the kill of a large prey animal has been concluded, if it has occurred in the open, as in a mountain meadow, the carcass is dragged to a hiding place in brush or timber. A big lion is capable of moving even an adult bull elk, or a horse. Usually, although not always, the belly is ripped open first. Lions are especially fond of blood, liver, heart, and entrails. After eating its fill, the animal scratches sticks, grass, or rocks over the remainder, covering it in a cache to which it will return several times. If the weather is cool and the meat stays in good shape, the animal may make trips for a week or more until it has cleaned up all edible parts. Even bones are stripped of meat by licking action of the rough tongue.

Of course not every cache is revisited. If game is abundant, a lion may forsake cached meat for the excitement and hot pungent taste of the blood from a fresh kill. The cats are unpredictable. Ordinarily an individual will kill

*During a scientific study of the species, this half-grown cougar cub had been pursued by hounds until it was treed. Next it was tagged, weighed, examined and released unharmed.*

only when hunger dictates. Now and then, however, one goes on a killing binge, downing several deer for which it has no need, or wiping out a band of domestic sheep apparently just because they're easy.

## Movements

Except for the wandering of a male seeking a female, practically all of the travel of the lion is concerned with keeping its belly full. A predator that insists on fresh meat cannot be casual about hunting. It may not gorge every day, but it must keep roaming, making its hunting circuit endlessly, if it hopes to sustain itself by its prowess as a hunter of other living creatures.

It is believed that most of a lion's hunting movements are nocturnal. However, this is probably another of those half-truths born of the difficulties of observation of retiring, secretive animals. Animal callers operating at night have now and then called lions. Animal callers even more often have had a lion respond in broad daylight. Most of the very occasional sightings of lions in the wild have been made in daytime, although some lions have been seen crossing roads or trails in vehicle lights at night and rarely one has come into a hunter's camp at night. Females are known to take youngsters on daytime learning-to-hunt expeditions.

Probably the truth is that a lion hunts whenever it feels like it, moving both day and night as the urge strikes. It does hunt much by sight, but it sees well in very dim light and it may instinctively know that stalking prey is easier at night. The rarity of sighting mountain lions in the wild may be because they do most of their moving and hunting nocturnally, lying up to rest by day.

The size of any individual lion's hunting ground is determined by the abundance of game within it, and also, especially in the north by the season. A traveling cat requires more food when the weather is cold, and in addition food may be scarcer. A cougar ranging in high mountains on the summer range of mule deer, for example, has to make adjustments when the deer herd moves lower to its winter range. In some instances the predators may follow the deer, in which case hunting may be easier because the winter range is smaller as a rule, and the deer more concentrated.

Probably any lion moves several miles each day. Various studies indicate that most individuals travel an average of 5 to 10 miles a day—or night—consistently. A female with young still in a den must restrict her hunting range, coming back within each twenty-four-hour period to the den. When the kittens begin to hunt with her, she is also moderately handicapped. She cannot roam at will. During the period of raising the young a female unquestionably combs her bailiwick intensively, and is forced to depend more on a variety of forage. Males are not so restricted and therefore usually roam wider, covering at times as much as 20 to 25 miles in a single night.

Like many predators, even down to the diminutive weasel, the lion hunts more or less in a large circle. There may be much wandering meanwhile. The entire hunting area may be 50 to 60 miles or more across. But each individual has staked out a domain and unless driven out stays within it, coming back time after time to the same places within it, and time after time making the same general

*Starting from a crouch and for a short distance, probably no North American big mammal could outrun the mountain lion. These cats make leaping bounds of 15 to 20 feet with ease and manage to bring down healthy fleeing deer in chases of a few hundred yards.*

hunting circuit. There are advantages for the lion in this system. The larger the hunting area, the less severe the pressure upon the prey animals.

After a kill, the cat may quit roaming for several days and stay in the vicinity, going back to feed at its cache. It has been estimated that 10 pounds or less of meat will fill up a large cougar. Trappers of long experience believe that certain lions will rest after gorging and not eat for several days, then go back to the kill for more. Travel and meal frequency are, of course, interdependent. In what is probably the best study ever done on mountain lions, *The Puma, Mysterious American Cat,* by Young and Goldman, published by the American Wildlife Institute some years ago, the authors relate the experience of a hunter in Colorado who was trailing a lion on its hunting route.

With dogs, he stayed on the trail for eleven days. During that time the lion, a female that when finally taken dressed out at 160 pounds, killed two adult bighorn rams and a buck deer. Whether other kills were made in remote places shortcut by the dogs was not known. After each ram kill, the lion apparently had rested for a couple of days, going back to feed until each was almost wholly eaten. The hunter was able to catch up to the lion after the deer was killed. His estimate was that this lion had made a kill and gorged on the fresh new one every three days. If most lions follow a similar pattern, that would total at least 122 kills annually!

The cougar is not migratory, although it might at times seem to be. When one shows up where none has been known before, it is undoubtedly for one of several reasons. Perhaps food has become scarce on its previous hunting ground; or its chief forage has moved, as in a mule deer winter migration; or possibly competition from other lions has caused it to seek a new home. Young lions must scatter as they grow to adulthood to establish their own roaming grounds, even though in some instances it is known that several, of the same sex or both sexes, occasionally hunt the same region and even feed communally.

Because the puma is a constant traveler as it goes its hunting rounds, and each animal is quite individual in habits, it is not possible to state precisely how much roaming it does as a species, or guess why certain lions make certain trips. For example, occasionally a lion will cross an open flat several miles wide, traveling from one mountain range to another. This belies its nature of keeping to cover. In southern Arizona one was discovered bedded and resting out the day in the shade of a single bush out in the middle of an open desert expanse between ranges.

When traveling on a hunt, a mountain lion walks, meandering, casting about for clues to game with all its senses. A low-to-the ground stalk may be swift, or awesomely patient. As mentioned previously, the big cat is amazingly swift for short bursts. Individuals have been known to catch an alerted and running deer within a couple hundred yards—and the deer itself is swift. But the lion cannot keep up such a pace. It is a fantastic jumper, bounding when pressed from ledge to ledge 15 to 20 feet with ease, and on occasion even farther. Game biologists who have built supposedly predator-proof fences around pastures containing such animals as desert bighorns for breeding stock—fences above 8 feet high, of net wire and electrified at the top—have found to their exasperation and enlightenment that a determined lion goes over the top with ease.

A lion ahead of dogs can leap to a tree limb 15 feet above ground, or leap out of a tree when need be from a height of 30 feet or more without damage. Like most of the cat family, the cougar dislikes getting into water, but it is an excellent swimmer when the need arises—and whimsically sometimes when there seems to be no need at all. Pushed by hounds, a lion will occasionally swim a stream or a pond. A few have been observed casually swimming large rivers without any pressure from hounds or hunters.

## Breeding

Some of the movements of the mountain lion, particularly of males, are concerned with the frequent search for females ready to be bred. With few exceptions pumas are solitary creatures, except for females with young, or occasional immature animals of like or differing sex ranging together. At the earliest they do not breed until they are two years old, and may not until they are three. There is no seasonal mating period. Though most young have been observed in spring, trappers in various parts of the range have recorded taking pregnant females around the entire year.

Mountain lions are therefore similar in breeding habits to domestic cats. The female has regular periods of heat. As soon as a litter of kittens is born, she is ready to be bred again. Observations in zoos have filled in much knowledge of such details. The heat period lasts for roughly nine days. Obviously, when a male picks up the scent of a female in breeding condition, he is diverted from his prey hunting and his travels are centered on finding a mate. These hunts for females ready to be bred cause trouble occasionally among males. If lions are fairly abundant on a given range, several may track the same female.

Statistics kept for many years of hundreds of lions killed by trappers and others in several states indicate that the sex ratio is close to even. In some locations females have been slightly more numerous than males, and vice versa. The 50-50 ratio is undoubtedly general. This is certain to mean that several males at a time will be aware of any female nearby that is in heat. Battles between males ensue. Most researchers doubt that many of these fights are violently severe or determined, although some may be. A few fatalities have been authenticated.

Whichever male wins the contest claims the female. Again, this is quite similar to the habits of domestic cats. However, it does not mean that the first male to breed the female is the only one. Later she may accept several other males, if any are in the territory. Or a male may stay with or near a female for several days, with frequent breeding. Possibly competition among breeding lions is heightened because many live rather long lives. Although the average life span is thought to be about seven years, numerous individuals both in zoos and in the wild have been known to more than double that.

During mating periods, it is fairly well agreed among scientists who have diligently pursued the matter, mountain lions utter the same general types of sound as domestic cats, except with much greater volume. Although lions are most of the time exceedingly quiet, nonvocal creatures, the matter of the "scream of the catamount" is likely to bring on heated argument among any group of hunters, predator trappers, or even zoo keepers who have substantial knowledge of mountain lions. For over two hundred years this argument has gone on. Modern outdoor magazines hardly ever get through a year without a pair of opposing I-was-there stories, one claiming the puma screams like an anguished, terribly frightened woman, the other just as adamant that it never utters such a sound.

Supposedly the hair-raising shrieks have frightened scores of campers, woodsmen, and dwellers in remote places—or so many have attested. Some of these sounds probably were not lions at all, but the wails of mating owls. Some others may have been imagined, or sworn to simply because it seemed like a dramatic tale. However, zoo personnel who have had daily contact with numerous mountain lions over many years agree that the females at least do "scream" during mating periods. The male may or may not do likewise, but it does utter a sharp whistle.

It is indeed curious that there should be such lack of thoroughly authenticated information about the screaming of pumas. Pet lions purr loudly when petted, exactly like housecats except louder. Cornered or treed lions sometimes growl, and invariably hiss and spit when closely approached. Reliable observers agree that cougars do "yowl" and "caterwaul" at various times, much in the manner of domestic cats.

## Birth and Development

Mountain lion kittens mew like any other kittens, and the mother reassures them with a low grunting sound. As they grow to playful size, when alarmed they sometimes utter a shrill whistle. Cougar kittens, like most baby animals, are extremely appealing little creatures, weighing a pound or less at birth. They are quite different in appearance from their solid-color parents. The coat is spotted, and the tail has ringlike markings. Most of the world's cats have spots or stripes as adults. It may be that the markings of mountain lion young are an indication of ancient ancestors that did not grow up to be of a solid color.

The kittens are born a few days more than three months from the time the female is bred. Like domestic kittens, at birth their eyes are closed, but begin opening after ten days and are fully open within a couple of weeks. They are born in a well-hidden place; the mother hopes to avoid intrusion by other lions, particularly males that might kill the young. The den may be just that, in a rocky cave or beneath a protected ledge. Or it may be under the roots of a big tree or beside a down log. In mild climates the hideaway is simply in the midst of an extremely dense thicket of brush.

It is doubtful that the female breeds every year. Scientists believe a litter every two or three years is normal. However, as previously mentioned, kittens may be born any month in the year. They are exceedingly playful, romping and tumbling with each other and grappling with the mother's tail or paws when she is in the den. They nurse the mother for at least a month, often longer, although she may try to wean them after four or five weeks. If she does not, even though they begin eating meat when roughly a month and a half of age, they will continue to nurse until they have half their growth. Few are allowed that privilege.

Average litters contain two kittens, with three, or one, not uncommon. Occasionally larger litters are born, with four, five, even six young. By the time the kittens are two months old they are usually as large as a big domestic cat, 8 or 10 pounds. But they do not grow even from babyhood very swiftly. At six months a young lion will weigh three or four times as much as at two months, averaging up to 35 pounds or more. Then it begins quickly to fill out and look more like an adult. When it is a year old if in good health it may weigh 60 or 70 pounds. Meanwhile, as it grows the markings on its coat become less and less distinct and finally disappear.

When the kittens are small and still in the den but ready to eat meat, the mother at first brings food to them, or in some instances urges them to follow her to the kill, if it is not too far. Then she returns them to the denning place. By the time they are fully weaned and well able to travel, at possibly six months of age, the youngsters are beginning to feel the hunting instinct on their own. The female may take only one on a hunt, and allow it to try a stalk and kill, although she may have to help in the kill. They all make foraging trips together as the youngsters grow. They learn from their mother, the trips become longer, and the den site is forgotten. When a kill is made, mother and youngsters all feed on it together.

As a family group, ties are fairly strong. If danger threatens, such as a hound pack on the trail, the female may make a stand to try to fight them off. But if the hunter following the dogs shows up, the cats usually scatter, running and treeing. Young mountain lions that begin playfully hunting each other as babies around the den site soon switch to using the stalking tactics they've been perfecting on prey. But they are not eager to leave the mother and go it alone until they are at least a year old. Un-

*Like all cats, cougar cubs are playful, but in truth spend little time with their mother until well grown. The mother is kept too busy hunting to feed all. Actually, little is known about family life of the secretive mountain lion.*

doubtedly they depend too much on her hunting abilities. Occasionally young lions twice that age still run with the mother. If they persist until she is ready to be bred again, she irritably chases them off, and each starts a life of his own.

## Senses

Like all cats, the mountain lion is equipped with exceptionally keen eyesight, because its livelihood depends upon its stalking ability. Although the big cat is colorblind, living therefore in a world of varying shades of gray, it sees extremely well in the dimmest light, in what to the human eye would be classed as total darkness. The eyes, which in bright light become mere vertical slits, are equipped with literally millions of light-gathering cells. In darkness the iris opens until the entire eye seems to be pupil. The eyes are slitted in bright light because of their extreme sensitivity. When a bright light is directed into the eyes of a lion at night, they glow brightly, as do those of all night-feeding animals, even deer. Conversely, the human eye, far less well supplied with light-gathering cells, does not glow in a light at night.

Stationary objects are difficult for the cat to identify. But it masterfully detects the slightest motion. However, in all animals it is difficult to tell at what point one sense is aided by another. For example, the sense of smell in the cougar is keen, but by no means as sharp as that of the trailing animals such as wolves and coyotes. At closer ranges than the wild dogs it uses scenting ability in conjuction with sight, and of course hearing. It may pick up the smell of a deer drifting downwind, move in until it can see the animal, then bring both senses to bear. Hearing is also sharp, and all three senses are utilized as a battery during any hunt

The lion also has most extremely sensitive whiskers, each with nerves at the base that relay messages. Thus a cougar creeping through brush in darkness "feels" for openings with its whiskers, slithers through narrow places without a sound because it is aware of the proximity of every twig. Unquestionably this acute sense of touch is an important asset to the stalk.

## Sign

Because the lion is so seldom seen, familiarity with its sign is especially important to the student of wildlife and to the hunter. Seeing the tracks of a puma in soft earth near a desert spring, or in dust at the mouth of a cave, or in snow high in the Rockies, is as close as most observers ever will get to the animal itself, and therefore they have high interest and significance.

It is not always easy for the casual visitor to wilderness areas to distinguish between the tracks of the wild dogs and wild cats. Except under unusual circumstances, cat tracks do not show the claw marks. The claws are kept retracted. Most wolf and coyote tracks have def-inite toenail imprints. The toe prints of the cat are usually more distinctly separated from the heel print than those of the dogs, and they also form a wider and more rounded arc out in front of the heel-pad mark. The heel print usually is in addition larger and much more distinct than that of the dogs.

The tracks of an adult cougar can hardly be confused with tracks of the smaller wild cats, such as bobcats and lynx. They are on the average at least 3 and as much as 4 to 4 ½ or more inches long for each foot, front and rear, with the print of the front pad wider than it is long, and also wider and somewhat larger than the print of the hind foot. Conceivably, in portions of Mexico and Central America, the tracks of lion and jaguar might be confused. However, an adult jaguar is a larger, heavier animal, with correspondingly larger track. In some locations, where snow is deep, a lion may leave marks of its dragging tail. This is a specialized sign and seldom observed.

Like all cats, the lion has a habit of covering its scats, scratching a slight depression, depositing the droppings, and covering them with scratched-up dirt or leaves and twigs. However, it may also simply deposit droppings on a rocky place with no attempt to cover them. "Scent posts" are common, and a sign lion hunters seek. These are small heaps of earth as a rule, scratched together, upon which the animal urinates. Another lion that comes upon this scrape will pause and also urinate on it, and sometimes claw up more dirt. In scientific studies, and in trapping, urine collected from a captive cougar has often been used on scratch spots to freshen them and draw other lions to them.

Now and then one may happen upon what has been a meat cache, or even upon one with parts of a kill still in it. Most of these are hidden too well, however, for hikers or casual travelers in lion country to find them. Droppings, though quite obvious and easily identified, also are not often discovered or recognized, except

*Unlike some other big-game animals, no single sense of a mountain lion seems to be dominant. Good sight, smell, and hearing interplay in bringing the big cat to its prey.*

by hunters and trappers seeking lion sign in a specific territory. They are large and are usually full of hair and of bits of bone from the kills. In form they may be large pellets of varied shapes, or a continuous, long scat with deep corrugations in it at unequal intervals. Droppings may be on or near the scent mounds, as well as covered elsewhere.

The lion, like the domestic cat and other wild cats, rakes its claws on tree trunks, presumably to sharpen them. Although a schooled trapper or hunter might spot this sign, it is by no means distinct; the scratches are not as deep and plain as those made, for example, by bears. Thus this sign is rarely noted by average observesrs. When a lion is walking, the average stride of each leg is about 2 feet. Quite often the tracks are blurred and confusing because the hind foot is brought forward and set down almost exactly into the impression of the front foot.

This instantly tags the puma as a stalker. The front foot is carefully placed to avoid any slightest sound. This is a safe place into which the following hind foot is instinctively set. When a lion track shows plainly for some yards, the direction is ordinarily in a fairly straight line, not meandering or dodging and darting from side to side.

## Hunting

Tracks and the scrapes of the mountain lion are what hunters seek for evidence that a lion is using a particular territory. There are only two methods of lion hunting that have even remote chances of success. The most common and generally most productive is with a pack of specially trained dogs. On occasion an excited and less than well-informed sportsman has come upon cougar tracks and decided to trail the animal down. Only by the rarest plain luck is such an endeavor successful.

The second method of lion hunting that has ended successfully in recent years for a scattering of outdoorsmen is calling. The call used is a standard coyote or "varmint" call, mimicking the squall of an injured rabbit. In order to have any chance at all of calling success, the hunter must be positive a lion is resident in the area to be hunted. The caller takes a stand where he can watch broken approaches, for example in a remote canyon, and begins blowing the raucous call. If a lion happens to be lying up within hearing distance, it may come to investigate. The thrill is tremendous when this does occur. However, of numerous callers who have tried this, only a very few have been successful. It is a gambler's game.

In almost all states and provinces today the hunting of the mountain lion is carefully controlled, with season limits, one-lion bag limits, and even stipulated areas that may or may not be hunted. In country where a substantial lion population is present, a few experienced lion hunters, some of them former or present government predator trappers, keep and train hound packs, and offer their services as guides to hunters who wish to book a hunt.

Thus there is really no hunting by individual sportsmen nowadays. It is just too expensive to keep and train the dogs. Most booked hunts are also quite expensive, and success, though chances are fair with an experienced outfitter, is far from certain. The sport is exceedingly rugged. Nonetheless, the chase is thrilling. Some sportsmen book lion hunts not to kill a lion but simply to enjoy the wild experience of the chase. Sometimes a lion that doesn't tree, or leaps out of a tree and is caught by dogs, swiftly kills several of them. It is capable of swatting and slashing to death a whole pack. Curiously, however, a mountain lion will run from even a small, yapping cur dog, unless it is a female with young. For some reason, with few exceptions the big cats have no pluck when it comes to standing ground before a dog.

*Lion hunter Willis Buttolph of Utah heads up a trail with his pack of top-notch lion hounds. Hunters seldom succeed in tracking a lion without hounds trained to tree it.*

Much has been written about the personality of the cougar as a stoic, and sometimes as a downright coward. "Coward" is hardly the word. Left alone in the wild, it is by no means one. Rather, it is a shy and supremely secretive creature without notable enemies, one that likes to be left alone. Only man and his dogs seem to unduly disturb this animal as true enemies. Stoical, on the other hand, the lion certainly is. Scores of instances are recorded of lions treed or trapped sitting calmly and without undue fuss seeming to accept their fate. Perhaps this is in some degree a reflection of the amazing patience and discipline the cat displays in its hunting.

Much has also been written over recent years decrying the hunting of the mountain lion, and the placing of it on the list of game animals. Those who pursue this tack, though perhaps sincere, are in error. Before the mountain lion won its place as a game animal in state after state, it was indiscriminately killed, usually with bounty. To stockmen it is, properly in the case of individual lions, intolerable.

The bounty, and the lack of protection as a game animal, long allowed continuous persecution of the big cats. Now in all but a few locations the mountain lion has come into its own as a game animal, a status that gives it haven. By the careful and meager cropping of surplus animals through trophy hunting, the lion population is kept in control, stock killing on wilderness-fringe ranches and public forest grazing lands is held to a minimum, and the great cat is able to maintain an unendangered status.

Further, in areas where the moutain lion is not abundant enough to allow token hunting—which is all it gets anywhere nowadays anyway—all hunting can be, and in most instances has been, stopped. This system is one that state and federal game-department personnel have fought for over many years. It assures that this large, handsome, and graceful cat of the wilderness places still left to us will continue to be a mysterious part of the native fauna of the continent, the basis for legend, argument, and fireside tales for as long as those wilderness retreats are still with us.